A TEXT BOOK OF ATOMIC FORCE MICROSCOPY

Bijender Bahl

CYBER TECH PUBLICATIONS
4264/3, Ansari Road, Daryaganj, New Delhi-110002 (India)
Ph.: 011-23244078, 011-43559448 Fax: 011-23280028
E-Mail: cyberpublicationsdelhi@yahoo.com
Website: www.cybertechpublications.com

A TEXT BOOK OF ATOMIC FORCE MICROSCOPY

Bijender Bahl

First Edition 2011

Published by :

G.S. Rawat for **Cyber Tech Publications**
4264/3, Ansari Road, Daryaganj, New Delhi-110002 (India)
Ph.: 011-23244078, 011-43559448 Fax: 011-23280028
E-Mail: cyberpublicationsdelhi@yahoo.com
Website: www.cybertechpublications.com

A TEXT BOOK OF ATOMIC FORCE MICROSCOPY

CONTENTS

Preface

Probably every child spends some time wondering why adults are so crazy. The question is usually answered in early adulthood with the realization that our parents' revenge is that we become them. Several decades ago microbiology spawned molecular biology. Brash and relentlessly reductionist in its youth, molecular biology seemed on the verge of solving every biological problem in sight. Biological problems did not vanish as molecular biology matured, and now the science is accused by those who do not understand it of being merely a collection of techniques. This is nonsense. Good biology-observation, discernment of pattern, testing of generalization-can be practiced quite effectively at the molecular level. This is appreciated full well by microbiologists, who depend on biochemistry and molecular biology in order to understand microorganisms. It is gratifying that molecular biologists increasingly appreciate the extraordinary diversity of microorganisms as a source of good biological questions. Microbiology also demands a sense of physiology, genetics, population biology, ecology, and evolution, so the field expands the minds of its practitioners. All biologists, indeed all scientists, are welcome in microbiology, a place where visitors tend to stay.

This book is primarily a textbook for lecturers and graduate and undergraduate students. To this group the book offers a thorough introduction to evolutionary computing (EC), including the basics of all traditional variants (evolution strategies, evolutionary programming, genetic algorithms, and genetic programming); EC themes of general interest (such as algorithm parameter control, or constraint handling); a collection of particular EC techniques (e.g., niching, or coevolution); and an outlook to related areas (evolutionary art). This book is also meant for those who wish to apply EC to a particular problem or within a given application area. To this group the book is valuable because it presents EC as something to be used, rather than just being studied, and it contains an explicit treatment of guidelines for good experimentation. Last, but not least, this book contains information on the current state of the art in a wide range of subjects that are interesting to fellow researchers as quick reference on subjects outside of their own specialist field of evolutionary computing.

The motivation behind the book is education oriented. Both authors have many years of teaching experience, that is, have taught EC many times, not only within the context of a university, but also at EC-related summer schools for doctoral students, and at commercial courses for practitioners from business and industry. The lack of one good textbook that covers all necessary aspects of EC, contains the factual knowledge but also paying attention to the skills needed to use this technology has been repeatedly experienced.

— *Author*

1

How the Atomic Force Microscope Works

INTRODUCTION

Microscopes have always been one of the essential instruments for research in the biomedical field. Radiation-based microscopes (such as the light microscope and the electron microscope) have become trustworthy companions in the laboratory and have contributed greatly to our scientific knowledge. However, although digital techniques in recent years have still enhanced their performance, the limits of their inherent capabilities have been progressively reached.

The advent of scanning probe microscopes and especially of the atomic force microscope has opened new perspectives in the investigation of biomedical specimens and induces to look again with rejuvenated excitement at what we can learn by "looking" at our samples. Novices are at first mesmerized by two features: the name of the instrument and the colorful 3D computer visualization of surfaces. One later learns that quite often it is not possible to obtain the "atomic" resolution that one hoped to achieve but that nevertheless images do contain details not observable with any other instrument.

The tri-dimensional mapping of the surface gains scientific relevance when one realizes that it is not just fancy surface reconstruction but that true topographic data with vertical resolution down to the subnanometer range is readily available. Moreover, when simplified sample preparation and the possibility of investigating specimens in liquid environment become apparent, one becomes convinced that he or she must find a way to apply AFM to his or her own research.

PERFORMANCE RANGE OF AFM

AFM images show significant information about surface features with unprecedented clarity. The AFM can examine any sufficiently rigid surface either in air or with the specimen immersed in a liquid. Recently developed instruments can allow temperature control of the sample, can be equipped with a closed chamber for environmental control, and can be mounted on an inverted microscope for simultaneous imaging through advanced optical techniques.

The field of view can vary from the atomic and molecular scale up to sizes larger than 125 μm so that data can be compared with other information obtained with lower resolution techniques. The AFM can also examine rough surfaces because its vertical range can be up to 8–10 μm. Large samples can be fitted directly in the microscope without cutting. With stand-alone instruments, any area on flat or nearly flat specimens can be investigated. In addition to its superior resolution with respect to optical microscopes, the AFM has these key advantages with respect to electron microscopes. Compared with the scanning electron microscope (SEM), the AFM provides superior topographic contrast, in addition to direct measurements of surface features providing quantitative height information.

Because the sample need not be electrically conductive, no metallic coating of the sample is required. Hence, no dehydration of the sample is necessary as with SEM, and samples may be imaged in their hydrated state. This eliminates the shrinkage of biofilm associated with imaging using SEM, yielding a nondestructive technique. The resolution of AFM is higher than that of environmental SEM, where hydrated images can also be obtained and extracellular polymeric substances may not be imaged.

Compared with transmission electron microscopes, 3D AFM images are obtained without expensive sample preparation and yield far more complete information than the 2D profiles available from cross-sectioned samples. In the following subheadings we will give a brief outline of how the AFM works followed by a description of the parts that can be added to the basic instrument. Our overview makes no pretense to completeness but aims at simplicity. For a more thorough description of the physical principles involved in the operation of these instruments, we refer you to the specialized literature.

The Microscope

In Fig. 1, a schematic diagram of an AFM is shown. In principle, AFM can bring to mind the record player, but it incorporates a number of refinements that enable it to achieve atomic-scale resolution, such as very sharp tips, flexible cantilevers, a sensitive deflection sensor, and high-resolution tip– sample positioning.

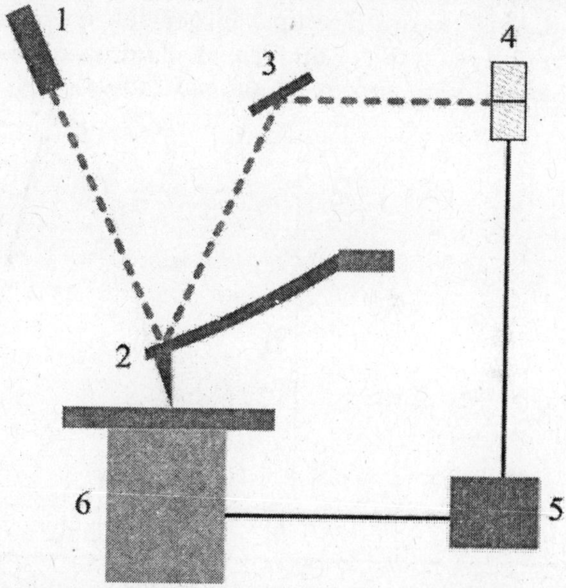

Fig. 1. Schematic diagram of a scanned-sample AFM. In the case of scanned probe, it is the tip that is scanned instead of the sample. 1, Laser diode; 2, cantilever; 3, mirror; 4, position-sensitive photodetector; 5, electronics; and 6, scanner with sample.

The Tip and Cantilever

The tip, which is mounted at the end of a small cantilever,

is the heart of the instrument because it is brought in closest contact with the sample and gives rise to the image though its force interactions with the surface. When the first AFM was made, a very small diamond fragment was carefully glued to one end of a tiny piece of gold foil. Today, the tip–cantilever assembly typically is fabricated from silicon or silicon nitride and, using technology similar to that applied to integrated circuit fabrication, allows a good uniformity of characteristics and reproducibility of results. The essential parameters are the sharpness of the apex, measured by the radius of curvature, and the aspect ratio of the whole tip.

Although it would seem that sharper tips should yield more detailed images, this may not occur with all samples: in fact, quite often, so-called "atomic resolution" on crystals is obtained best with standard silicon nitride tips. In general, one can choose among one of three types of tip. The standard tip is usually a 3-μm tall pyramid with approx 30-nm end radius.

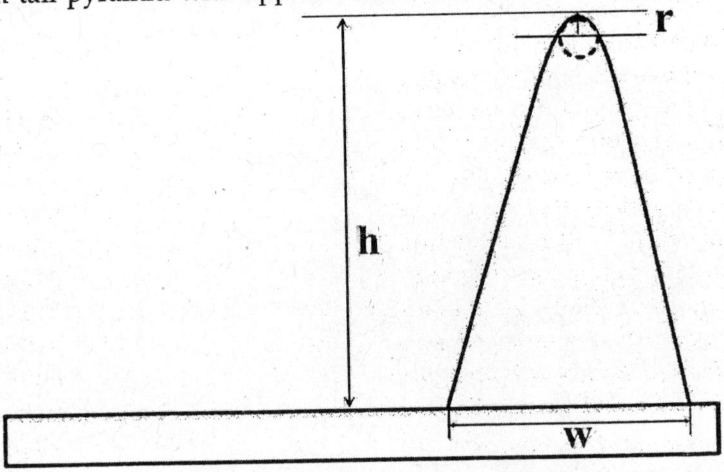

Fig. 2. The essential parameters in a tip are the radius of curvature (r) and the aspect ratio (ratio of h *to* w*).*

The electron-beam-deposited tip or "super tip" improves on this with an electron-beam–induced deposit of material at the apex of the tip, offering a higher aspect ratio and end radius than the normal tip, albeit with the drawback of fragility. Finally, tips made from silicon (either polysilicon or single crystal) through improvedmicrolithographic techniques have a higher aspect ratio and small apex radius of curvature, maintaining reproducibility and durability.

The cantilever carrying the tip is attached to a small glass "chip" that allows easy handling and positioning in the instrument. There are essentially two designs for cantilevers, the "V" shaped and the single-arm kind, which have different torsional properties.

The length, width, and thickness of the beam(s) determine the mechanical properties of the cantilever and have to be chosen depending on mode of operation needed and on the sample to be investigated. Cantilevers are essentially classified by their force (or spring) constant and resonance frequency: soft and low-resonance frequency cantilevers are more suitable for imaging in contact and resonance mode in liquid, whereas stiff and high-resonance frequency cantilevers are more appropriate for resonance mode in air.

Deflection Sensor

AFMs can generally measure the vertical deflection of the cantilever with picometer resolution. To achieve this, most AFMs today use the optical lever or beam-bounce method, a device that achieves resolution comparable to an interferometer while remaining inexpensive and easy to use.

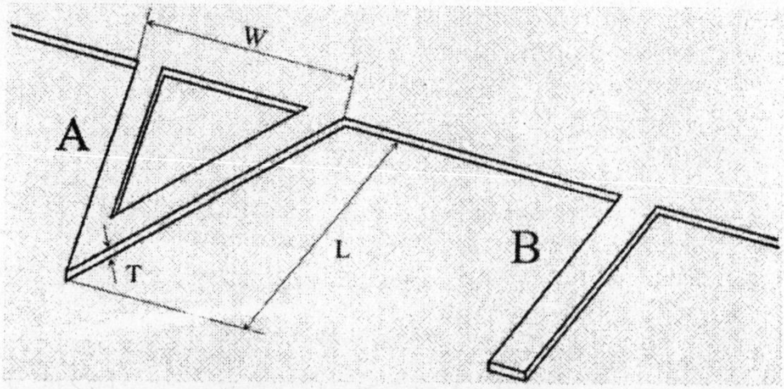

Fig. 3. Triangular (A) and single-beam (B) cantilevers. The mechanical properties, such as the force constant and resonant frequency, depend on the values of width (W), length (L), and thickness (T).

In this system, a laser beam is reflected from the backside of the cantilever (often coated by a thin metal layer to make a mirror) onto a position-sensitive photodetector consisting of two side-by-side photodiodes.

In this arrangement, a small deflection of the cantilever will tilt the reflected beam and change the position of beam on the photodetector. The difference between the two photo-diode signals indicates the position of the laser spot on the detector and thus the angular deflection of the cantilever.

Because the distance between cantilever and detector is generally three orders of magnitude greater than the length of the cantilever (millimeters compared to micrometers), the optical lever greatly magnifies motions of the tip giving rise to an extremely high sensitivity.

Image Formation

Images are formed by recording the effects of the interaction forces between tip and surface as the cantilever is scanned over the sample. The scanner and the electronic feedback circuit, together with sample, cantilever, and optical lever form a feedback loop set up for the purpose. The presence of a feedback loop is a key difference between AFM and older stylus-based instruments so that AFM not only measures the force on the sample but also controls it, allowing acquisition of images at very low tip-to-sample forces.

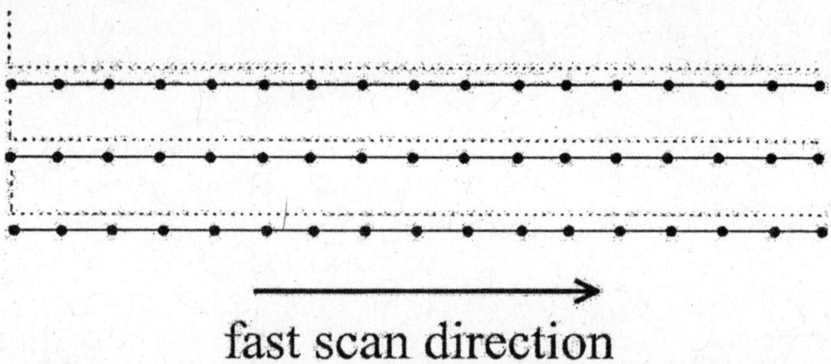

Fig. 4. Raster scan for image acquisition. The AFM electronics drive the scanner across the first line of the scan and back. The scanner then steps in the perpendicular direction to the second scan line, moves across it and back, then to the third line, and so forth.

The scanner is an extremely accurate positioning stage used to move the tip over the sample (or the sample under the tip) to form an image, and generally in modern instruments is made from a piezoelectric tube. The AFM electronics drives the scanner across the first line of the scan and back. It then steps in the perpendicular direction to the second scan line, moves across it and back, then to the third line, and so forth.

As the probe is scanned over the surface, a topographic image is obtained storing the vertical control signals sent by the feedback circuit to the scanner moving it up and down to follow the surface morphology while keeping the interaction forces constant. The image data are sampled digitally at equally spaced intervals, generally from 64 up to 2048 points per line. The number of lines is usually chosen to be equal to the number of data points per line, obtaining at the end a square grid of data points each corresponding to the relative x, y, and z coordinates in space of the sample surface.

Usually during scanning data are represented by gray scale images, in which the brightness of points can range from black to white across 256 levels corresponding to the information acquired by the microscope (that can be height, force, phase, and so on).

A VARIETY OF INSTRUMENTS AND OPTIONS

The first instruments introduced on the market had all very similar features and range of applications: they had scanners with small range, limited optical access, and could accommodate only small samples. Essentially they where built to make very high-resolution imaging on flat samples in a dry environment. As the possibilities of AFM were developed, a wider range of instruments, optimized for specific applications, have been developed. We can now find instruments that are specifically designed for large samples, such as silicon wafers, that have metrological capabilities, utilize scanner close loop operation, are optimized for liquid and electrochemistry operation, and can be mounted on an inverted microscope for biological investigations. Usually, one single instrument can have different options to extend its capabilities, but to date it is not possible to have an instrument that covers all possible applications with maximum performance. For this reason, it is necessary to have clearly in mind what will be the main features that are desired in an instrument before its purchase, understanding at the same time that a loss of performance in other aspects may be possible.

One can distinguish between two main classes: scanned-sample and scanned-tip microscopes. We give a brief description of the advantages of one system with respect to the other.

Scanned Sample

This scanned-sample AFM is the first design in which the sample is attached to the scanner and moved under the tip. Depending on how the cantilever holder, laser, and photodetector are assembled, it can easily accommodate an overhead microscope provided that long focal length objectives are used.

A clear view of where the tip is landing is usually possible, speeding up the time it takes to get a meaningful image of the sample.

Scanners with wide *x,y*, and *z* range are usually available and closed loop control feedback is more easily implemented in this scheme and often a lower mechanical noise level can be obtained allowing higher ultimate resolution.

There are quite a few drawbacks. First of all, the size and weight of the sample has to be limited because it is sitting on the scanner and may change its behavior.

For the same reason, operation in liquid is impaired because liquid cells tend to be small and difficult to seal, and liquid flow or temperature control are more complicated to implement. Notwithstanding these difficulties, excellent results can be obtained on typical biomedical science specimens by ingeniously adapting them to the instruments characteristics.

Scanned Tip

In the scanned-tip method of operation, the sample stays still and it is the cantilever, attached to the scanner, which is moved across the surface. Although for scanning tunneling microscopes this was one of the first solutions applied, to build a scanned tip AFM requires overcoming some difficulties, essentially related to adapting the beam bounce detection scheme to a moving cantilever. For this reason, it has been only recently that models made according to this design have been marketed, after appropriate technology was developed.

The first examples were the so-called "stand-alone" systems, usually an AFM resting on three legs and able to scan the surface of any object under its probe. Later, specialized instruments were developed, capable of being coupled or even integrated into inverted optical microscopes for biological applications.

With respect to the scanned-sample models, scanned-tip instruments can be more easily equipped with temperature-controlled stages, open or closed liquid cells, liquid flow systems, electrochemistry cells, and controlled atmosphere chambers. Concerning limitations, one could say that what is gained on one side is lost on the other.

For example, often the overall noise level is higher, limiting ultimate resolution. Large scan areas are more difficult to scan because tracking systems have to be used to keep the laser spot on the back of the cantilever. A top view of samples is obstructed by the scanner assembly: special hollow tubes have been developed recently, but even so on-axis microscopes, which are useful on nontransparent samples, will still have limited resolution and lateral field of view.

LOADING A SAMPLE IN THE MICROSCOPE

Imaging Dry Samples

Samples to be imaged in atmospheric environment are often simply glued to a sample holder, usually a metal disk. The disk is then inserted in the AFM, where it is held firmly by a small magnet.

An essential point is that the sample has to be firmly adherent to the sample holder; otherwise, very poor imaging will be achieved. For this reason, one has to be careful in the choice of the glue or sticky tape: slow drying glue or thick sticky tape should be avoided. A drawback is that after use in the AFM, the sample is difficult to take off without damage.

Some systems, usually scanned-tip, can accept samples directly, securing them with a metal clip or springs. This method allows sample recovery without damage for further use in other experiments, but it can be less stable and needs special care for high-resolution work. Sometimes, because of the ease of use of the AFM, one forgets to be careful while handling the sample and either fingerprints or dust from a dirty environment contaminates the sample. It is best to keep a reserved area of the laboratory free from contaminants for the operations of sample and cantilever mounting.

Imaging in Liquid

One of the main reasons for the success of AFM in biomedical investigations is its ability to scan samples in physiological condition, that is, immersed in liquid solutions. Just to make an example, scanned-tip systems can often be directly used to image cells into a standard Petri dish.

Each manufacturer has its own design of liquid cells, sometimes different ones depending on the application, and users may decide to make their own to fit specific needs. A few additional things that have to be taken care of when imaging in liquid are the temperature of the solution (eventually added during imaging) and maintenance of the liquid cell and cantilever holder assembly.

Because the cantilever is extremely sensitive to temperature changes, it is important to let the system equilibrate before taking images. For example, in the case of contact mode imaging with silicon nitride cantilevers and tips, a large change in time of the signal on the photodetector corresponding to cantilever deflection can be observed in the presence of a temperature change. If temperature is not stable prior to approach of the tip to the sample and one starts taking images, after some time the applied force could be quite different than at the beginning of the imaging session.

Once finished using the microscope for imaging in liquid, it is essential to immediately clean thoroughly all parts that have been in contact with the solution to avoid contamination of future experiments. Usually, it should be possible to disassemble and sonicate all vital parts of the liquid cell and the cantilever holder.

FUTURE DEVELOPMENTS

The AFM is part of a family of scanning probe microscopes that has a great growth potential. It is a fact that the majority of novel applications and techniques developed in scanning probe microscopes in the last years are related to the life sciences. There is still much room for technical improvement: electronics, scanners, and tips are constantly improving. Scan speed limitations, sample accessibility, and ease of use have been addressed and can be still improved. As more and more biomedical researchers will be involved in the use of AFM, with their experience they will be able contribute in developing an instrument less related to the physical science (its origin) and more tailored to our specific needs.

2

Imaging Methods in Atomic Force Microscopy

INTRODUCTION

One can easily distinguish between two general modes of operation of the atomic force microscope (AFM) depending on absence or presence in the instrumentation of an additional device that forces the cantilever to oscillate in the proximity of its resonant frequency.

The first case is usually called static mode, or DC mode, because it records the static deflection of the cantilever, whereas the second takes a variety of names (some patented) among which we may point out the resonant or AC mode.

In this case, the feedback loop will try to keep at a set value not the deflection but the amplitude of the oscillation of the cantilever while scanning the surface. To do this, additional electronics are necessary in the detection circuit, such as a lock-in or a phase-locked loop amplifier, and also in the cantilever holder to induce the oscillatory excitation.

From a physical point of view, one can make a distinction between the two modes depending on the sign of the forces involved in the interaction between tip and sample, that is, by whether the forces there are attractive or repulsive. In Fig. 1, an idealized plot of the forces between tip and sample is shown, highlighting where typical imaging modes operate. In the following we briefly describe the DC and AC modes of operation relevant to the kind of samples that usually are investigated in the biomedical field.

DC MODES

Contact Mode

Also called constant force mode, the contact mode is the most direct AFM mode, where the tip is brought in contact with the surface and the cantilever deflection is kept constant during scanning by the feedback loop.

Image contrast depends on the applied force, which again depends on the cantilever spring constant. Softer cantilevers are used for softer samples. It can be used easily also in liquids, allowing a considerable reduction of capillary forces between tip and sample and, hence, damage to the surface. Because the tip is permanently in contact with the surface while scanning, a considerable shear force can be generated, causing damage to the sample, especially on very soft specimens like biomolecules or living cells.

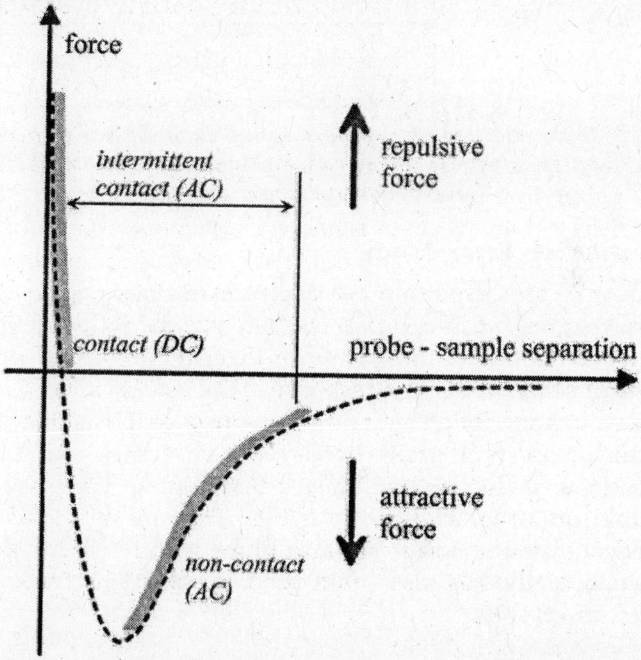

Fig. 1. Idealized plot of the forces between tip and sample, highlighting where typical imaging modes are operative.

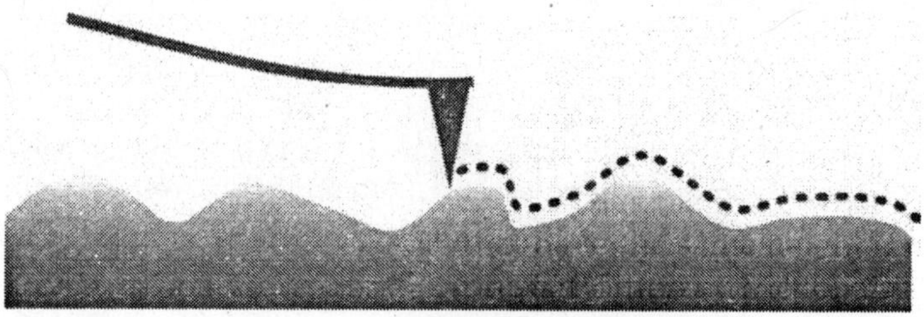

Fig. 2. In contact mode, the tip follows directly the topography of the surface while it is scanned.

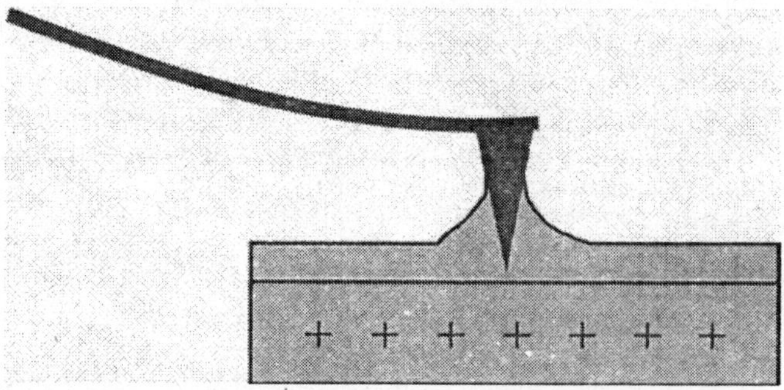

Fig. 3. In contact mode, capillary forces caused by a thin water layer and electrostatic forces can considerably increase the total force between sample and tip.

Deflection or Error Mode

In same cases, especially on rough and relatively rigid samples, the error signal (i.e., the difference between the set point and the effective deflection of the cantilever that occurs during scanning as a result of the finite time response of the feedback loop) is used to record images. By turning down on purpose the feedback gain, the cantilever will press harder on asperities and less on depressions, giving rise to images that contain high-frequency information otherwise not visible. This method has been extensively used to image submembrane features in living cells. The same method is also often used to record high-resolution images on crystals.

Lateral Force Microscopy

In this case (a variation of standard contact mode), while scanning the sample not only the vertical deflection of the cantilever but also the lateral deflection (torsion) is measured by the photodetector assembly, which in this case will have four photodiodes instead of two. The degree of torsion of the cantilever supporting the probe is a relative measure of surface friction caused by the lateral force exerted on the scanning probe. This method has been used to discriminate between areas of the sample that have the same height (i.e., that are on a same plane) but that present different frictional properties because of absorbates.

AC MODES

All AC modes require setting the cantilever in oscillation using an additional driving signal. This can be accomplished by driving the cantilever with a piezoelectric motor (acoustic mode) or, as developed more recently, by directly driving by external coils a probe coated with a magnetic layer (magnetic mode).

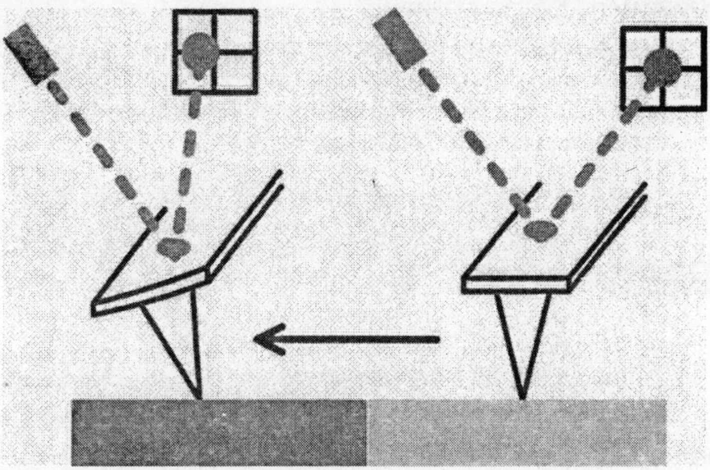

***Fig. 4.** Using a four-section photodetector, it is possible to measure also the torsion of the cantilever during contact mode AFM scanning. The torsion of the cantilever reflects changes in the surface chemical composition.*

This second method is giving interesting results, especially in liquid, as it allows better control of the oscillation dynamics and has inherently less noise.

Noncontact Mode

An oscillating probe is brought into proximity of (but without touching) the surface of the sample and senses the van der Waals attractive forces that induce a frequency shift in the resonant frequency of a stiff cantilever. Images are taken by keeping a constant frequency shift during scanning, and usually this is performed by monitoring the amplitude of the cantilever oscillation at a fixed frequency and feeding the corresponding value to the feedback loop exactly as for the DC modes. The tip–sample interactions are very small in noncontact mode, and good vertical resolution can be achieved, whereas lateral resolution is lower than in other operating modes. The greatest drawback is that it cannot be used in liquid environment, only on dry samples. Also, even on dry samples, if a thick contamination or water layer is present the tip can sometimes be trapped, not having sufficient energy to detach from the sample because of the small amplitude of oscillation.

Intermittent Contact Mode

The general scheme is similar to that of noncontact mode, but in this case during oscillation the tip is brought into contact with the sample surface so that a dampening of the cantilever oscillation amplitude is induced by the same

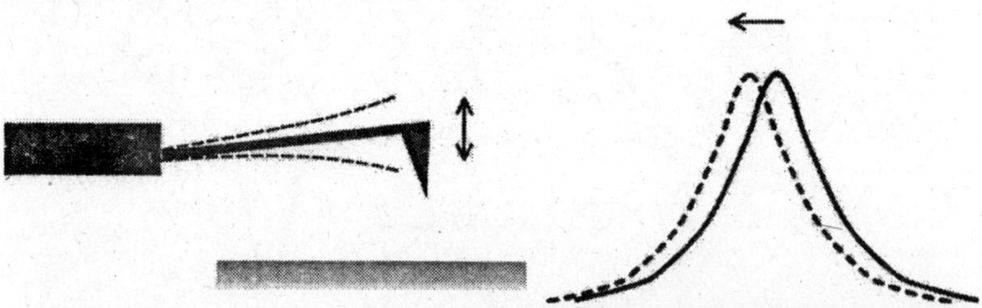

Fig. 5. In noncontact mode of operation, a vibrating tip is brought near the sample surface, sensing the attractive forces. This induces a frequency shift in the resonance peak of the cantilever that is used to operate the feedback.

repulsive forces that are present in contact mode. Usually in intermittent contact the oscillation amplitude of the cantilever is larger than the one used for noncontact. There are several advantages that have made this mode of operation quite popular. The vertical resolution is very good together with lateral resolution, there is less interaction with the sample compared with contact mode (especially lateral forces are greatly reduced), and it can be used in liquid environment. This mode of operation is the most generally used for imaging biological samples and is still under constant improvement, thanks to additional features such as Q-control or magnetically driven tips.

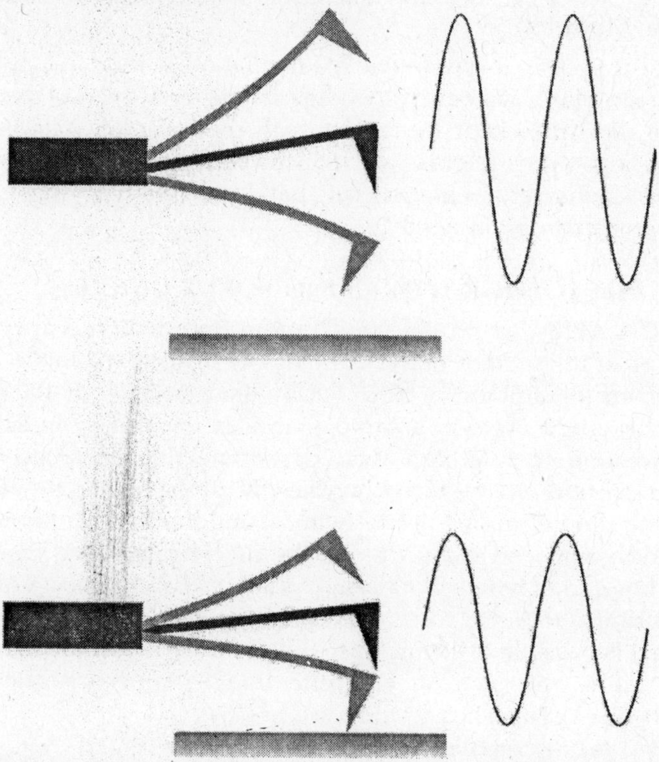

Fig. 6. In intermittent contact mode, the free oscillation of a vibrating cantilever is dampened when the tip touches the sample surface at each cycle. The image is performed keeping constant the oscillation amplitude decrease while scanning.

Phase Imaging Mode

If the phase lag of the cantilever oscillation relative to driving signal is recorded in a second acquisition channel during imaging in intermittent contact mode, noteworthy information on local properties, such as stiffness, viscosity, and adhesion, can be detected that are not revealed by other AFM techniques. In fact, it is good practice to always acquire simultaneously both the amplitude and phase signals during intermittent contact operation, as the physical information is entwined and all the data is necessary to interpret the images obtained.

Force Modulation

In this case, a low-frequency oscillation is induced (usually to the sample) and the corresponding cantilever deflection recorded while the tip is kept in contact with the sample. The varying stiffness of surface features will induce a corresponding dampening of the cantilever oscillation, so that local relative visco-elastic properties can be imaged.

BEYOND TOPOGRAPHY USING FORCE CURVES

The AFM can provide much more information than taking images of the surface of the sample. The instrument can be used to record the amount of force felt by the cantilever as the probe tip is brought close to a sample surface, eventually indent the surface and then pulled away. By doing this, the long-range attractive or repulsive forces between the probe tip and the sample surface can be studied, local chemical and mechanical properties like adhesion and elasticity may be investigated, and even the bonding forces between molecules may be directly measured. By acquiring a series of force curves, one at each point of a square grid, it is possible to acquire a so called force-vs-volume map that will allow the user to compute images representing local mechanical properties of the sample observed.

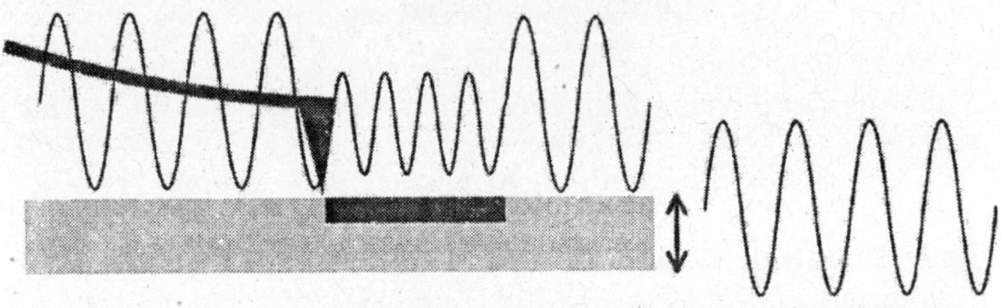

Fig. 7. During force modulation, the tip is kept in contact with the sample and the different local properties of the sample will be reflected in the amplitude of the oscillation induced in the cantilever.

Force curves typically show the deflection of the cantilever as the probe is brought vertically towards and then away from the sample surface using the vertical motion of the scanner driven by a triangular wave. By controlling the amplitude and frequency of the vertical movement of the scanner it is possible to change the distance and speed that the AFM probe travels during the force measurement. Conceptually what happens during a force curve is not much different from what happens between tip and sample during intermittent contact imaging. The differences are in the frequency, much lower for force curves, and the distance of travel of the probe, much smaller in intermittent contact. In a force curve, many data points are acquired during the motion, so that very small forces can be detected and interpreted by fitting the force curve according to theoretical models.

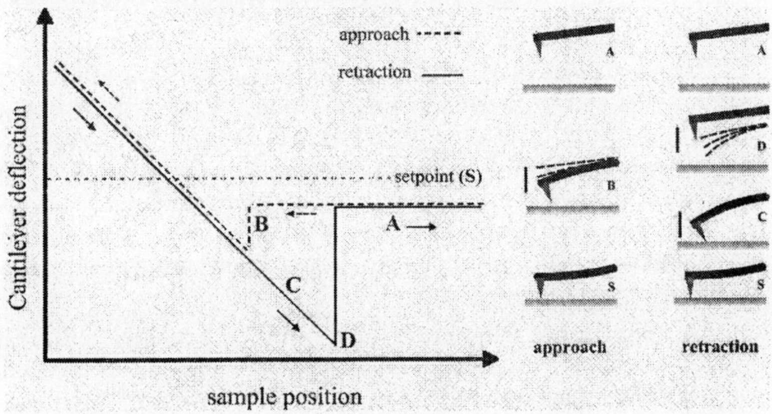

Fig. 8. Idealized force curve and cantilever behavior. From positions A to B, the tip is approaching the surface, and at position B contact is made (if an attractive or repulsive force is active before contact, the portion of the force curve will reflect it). After B, the cantilever bends until it reaches the specified force limit that is to be applied (S). Depending on the relative stiffness of the cantilever with respect to the sample, during this portion of the curve the tip can indent the surface. The tip is then withdrawn towards positions C and D. At position D, under application of the retraction force, the tip detaches from the sample (often referred to as 'snap off'). Between positions D and A, the cantilever returns to its resting position and is ready for another measurement.

Two details of technique are worth special care when obtaining quantitative data from force-vs-distance curves. The position-sensitive photodetector signal has to be calibrated so to measure accurately the deflection of the cantilever, and after calibration it is essential that the laser alignment is left unchanged. Usually the software of the AFM has a routine for such calibration, performed by taking a force curve on a hard sample and using the scanner's vertical movement as reference (which means that the scanner also has to be accurately calibrated). At this point, the curve we are plotting is not yet a force curve but a calibrated deflection curve. The next step is to convert it to a force curve using the force constant of the cantilever we are using. Manufacturers usually specify this value, but for each cantilever there can be quite large variations, so that for accurate work direct determination becomes necessary. There are different ways to measure the force constant, some requiring external equipment for measuring resonant frequency (such as spectrum analyzers) and others making use of reference cantilevers.

Form the point of view of biomedical applications, interesting experiments can be performed by coating the tip with a ligand and approaching through a force curve a surface where receptor molecules can be found. In this case the portion of the curve before snap off will have a different shape, reflecting the elongation of the bond between ligand and receptor before dissociation: from the shape the curve, it is possible to derive quantitative information on the binding forces.

If a force curve is taken at each point of a $N \times N$ grid, it is possible to derive images that are directly correlated to a physical property of the surface of the sample. For example, if the approach portion of each curve after contact is fitted using indentation theory, a map of the sample stiffness can be calculated. This data can be represented by an image in which the level of gray of each pixel, instead of representing the height of the sample, will correspond to the elasticity modulus. Similar images can be calculated for adhesion, binding, electrostatic forces, and so on.

3

Recognizing and Avoiding Artifacts in AFM Imaging

INTRODUCTION

Images taken with the atomic force microscope (AFM) originate in physical interactions that are totally different from those used for image formation in conventional light and electron microscopy. One of the effects is that a new series of artifacts can appear in images that may not be readily recognized by users accustomed to conventional microscopy. Because we are addressing ourselves to novices in this field, we would like to give an idea of what can happen while taking images with the AFM, how one can recognize the source of the artifact, and then try to avoid it or minimize it. Essentially, one can identify the following sources of artifacts in AFM images: the tip, the scanner, vibrations, the feedback circuit, and image-processing software.

TIP ARTIFACTS

The geometrical shape of the tip being used will always affect the AFM images taken with it. Quite intuitively, as long as the tip is much sharper than the feature under observation, the profile will resemble closely its true shape. Depending on the lateral size and height of the feature to be imaged, both the sharpness of the apex and the sidewall angle of the tip will become important. In general, the height of the features is not affected by the tip shape and is reproduced accurately, whereas the greatest artifacts are evident on the lateral geometry of objects, especially if they have steep sides. Avoiding artifacts from tips is achieved by using the optimal probe for the application: the smaller the size of the object, the sharper the tip. A notable exception arises in the case of high-resolution imaging on ordered crystals, where often better images are obtained with standard tips.

This can be explained by realizing that at this dimensional scale the measurable radius of curvature of the tip is not in fact involved in the imaging process, but instead smaller local protrusions on the apex of the probe will be the real tip (or tips) effectively taking the image.

Further understanding of AFM tip properties and related artifacts can be gathered from the vast literature on the subject, together with a variety of methods for their correction. Specific artifacts, depending on the mode of operation, have been investigated and explanations have been proposed.

Because we are now interested in showing a general overview of the subject for beginners in the field, we shall have a look at the main tip artifacts in a very simple way.

Features Protruding on the Surface Appear Larger Than Expected

In Fig. 1, the different profiles were obtained using a dull or a sharp tip when scanning a surface feature. In addition to sharpness, the geometrical shape also is important: a conical tip will affect the lateral shape of the feature less than a pyramidal one. Very small features, such as nanoparticles, nanotubes, globular proteins, and DNA strands, will always be subject to image broadening, so that the measured lateral size should be taken as an upper limit for the true size. Note that in all these cases the height of the sample will be reported accurately.

Repetitive Abnormal Patterns in an Image

When the size of the features on a flat surface is significantly smaller than the tip, repetitive patterns may appear in an image. Spherical nanoparticles or small proteins may assume an elongated or triangular shape reflecting the geometry of the apex of the tip.

Sometimes a so-called "double image" will appear along the fast scanning direction as a result of the presence on the tip of more than one protrusion slightly separated from one another and making contact with the sample (Fig. 2).

Pits and Holes in the Image Appear Smaller and Shallower

When the tip has to go into a feature that is below the surface, such as a hole, the lateral size and depth can appear too small and the tip may not reach the bottom. The geometry of the probe will dominate the geometry of the sample as is apparent from the line profile shown in Fig. 3. However, it is still possible to measure the opening of the hole from this type of image. Also, the pitch of repeating patterns can be accurately measured with probes that do not reach the bottom of the features being imaged.

Damaged or Contaminated Tips

If the probe is badly damaged or has been contaminated by debris from a less-than-clean sample surface, strangely shaped objects may be observed in the image and difficult to explain. For example, a damaged tip following the geometry of a regular test pattern (as in Fig. 4) will produce an asymmetric profile. In the case of contaminants, one often notices an abrupt change of detail contrast during scanning and a blurring of the image. Sometimes the debris particle may partially detach and is dragged along during scanning, leaving a diagonal track on the image that could be erroneously interpreted as a surface feature. Telltale signs in this case are the instabilities and glitches in the feedback signal that occur each time the particle is dragged along.

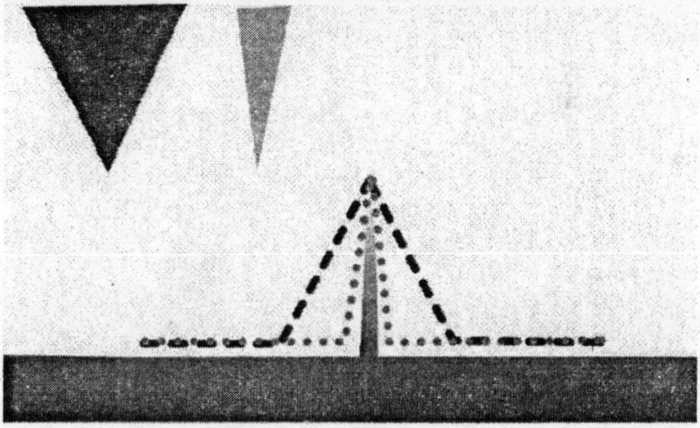

Fig. 1. Traces followed by a dull and a sharp probe as they go over a protruding feature. In such a measurement, the side of the tip will cause a broadening of objects in the image.

Fig. 2. A double tip will cause a shadow or double image along the scanning direction

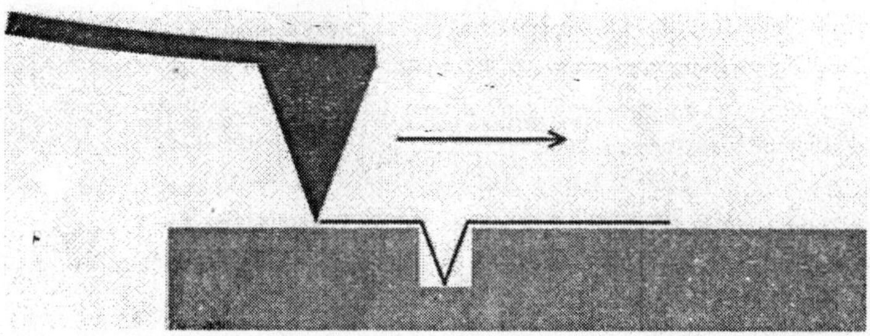

Fig. 3. Because of the width of the tip, the hole will not be faithfully reproduced.

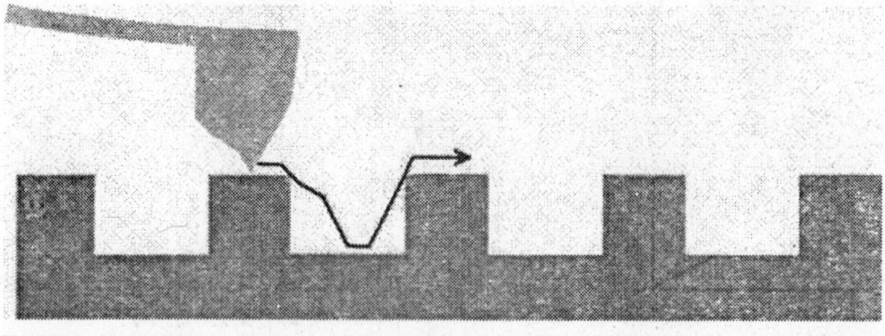

Fig. 4. A badly damaged tip creates artifacts while scanning a regular test pattern.

SCANNER ARTIFACTS

Piezoelectric ceramic scanners were one of the breakthroughs that made AFM possible. Their design has been constantly improved, but a number of artifacts still arise from their physical and mechanical properties. One point that must not be neglected is that scanner properties change with time and use. In fact, the piezoelectric material will change its sensitivity to driving signals if it is often used (it will become slightly more sensitive) or if it is left idle (it will depolarize and become less sensitive). The best thing to do is to periodically calibrate the scanner following the manufacturer's instructions.

Effects of Intrinsic Nonlinearity

If the extension of the scanner in any one direction is plotted as a function of the driving signal, the plot will not be a straight line but a curve similar to the one shown in Fig. 5. The nonlinearity may be expressed as a percentage (describing the deviation from linear behavior), and it typically ranges from 2–25%, depending on the driving signal applied and scanner construction. The effects will be present both in the plane and in the vertical direction.

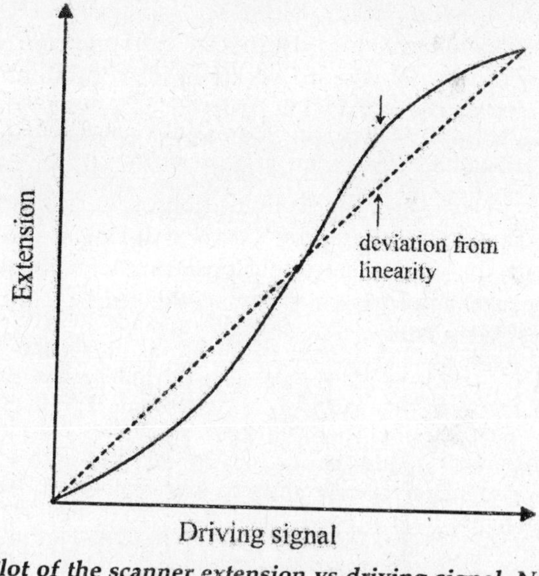

Fig. 5. Plot of the scanner extension vs driving signal. Notice the large deviation from linearity.

In the Plane

An AFM image of a calibration grid with periodic structures such as squares will appear severely distorted, with nonuniform spacing and curvature of features, typically appearing smaller on one side of the image than on the other (Fig. 6). On a generic sample with no regular pattern the distortion may not be recognizable, but it will be certainly present. Once the scanner is properly linearized, it is also critical that the scanner be calibrated. For example, it is possible for the scanner to be linear but not calibrated. If the calibration is incorrect, then the x and y values measured from line profiles will be incorrect.

In Plane Linearization

There are essentially two methods to linearize a scanner in the x and y directions: by software or hardware. Software correction is performed by mathematically modeling the nonlinear behavior of the scanner, finding the parameters for a correction algorithm imaging a known grid, and then applying the algorithm during scanning using the parameters stored in a look-up table.

The limits of this method lie in the fact that unfortunately the corrections strongly depend upon the scan speed, scan direction, and offset that have been used during the calibration procedure. When images in normal use are taken under conditions similar to the calibration, the correction will be accurate; otherwise, nonlinearities will be again present. More recently hardware correction for large scanners has become popular because it gives better results. In this case, the true position of the scanner in the x and y directions is measured by a sensor during scanning and compared with the intended scanner position. A feedback circuit applies an appropriate driving signal to the scanner in order to attain the desired position.

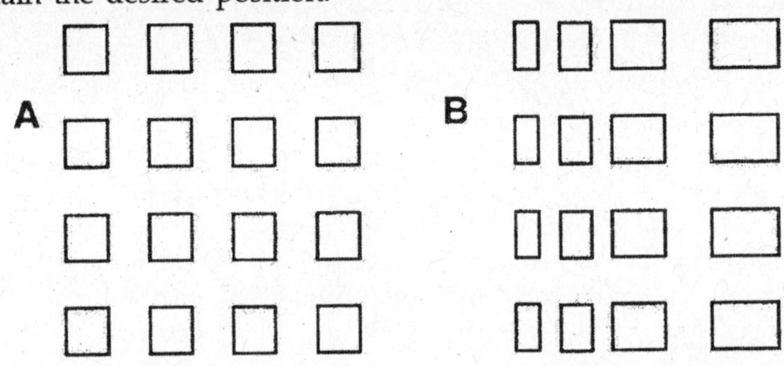

Fig. 6. Distortion of a test pattern caused by scanner nonlinearity.

In Height Measurements

Because the height range of scanners is usually an order of magnitude smaller than the range in the scanning plane, effects of nonlinearity are less severe but still present. To make accurate height measurements with an AFM, it is necessary to calibrate the scanner in the *z*-axis. Often the microscope is calibrated at only one height.

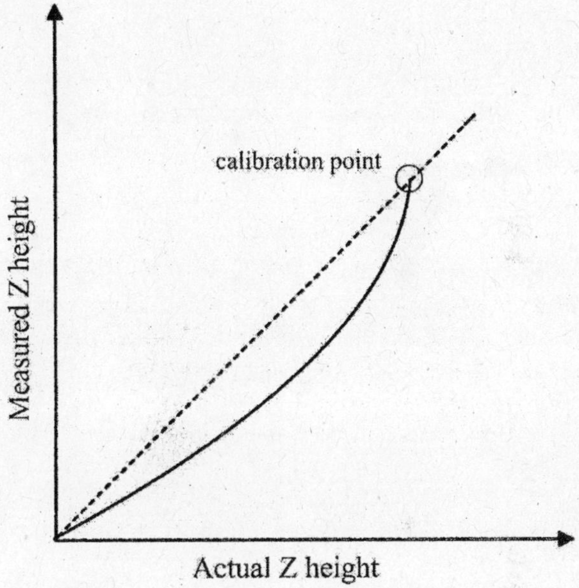

Fig. 7. Quite often, the z height response of the scanner is calibrated in only one point. The plot represents the deviation from the true value for measurement of heights that differ from the one at which the scanner has been calibrated.

This means that if the relationship between the measured z height and the actual z height is not linear, then the height measurements will not be correct unless the feature being observed has a height close to the calibration measurement (Fig. 7). It is also to be noted that although calibration gratings are reasonably easy to make by lithographic techniques, step–height calibration standards are more difficult to obtain, especially for very high-resolution work. Often researchers make their one reproducible height standards for accurate measurements in this range from crystals that have known height steps.

Effects of Hysteresis

All piezoelectric ceramics display hysteretic behavior, that is, if slowly scanned back and forth cyclically, to the same driving signal does not correspond the same position in the two scanning directions. This can be easily observed by comparing the profiles taken from left to right and in the opposite direction on a feature on the surface of a sample. The result would be like Fig. 8, where there is a lateral shift between the two profiles. Notice that an effect is also present in the vertical direction because the contraction and extension response of the scanner to the driving signal will be different, giving rise to an asymmetric step height.

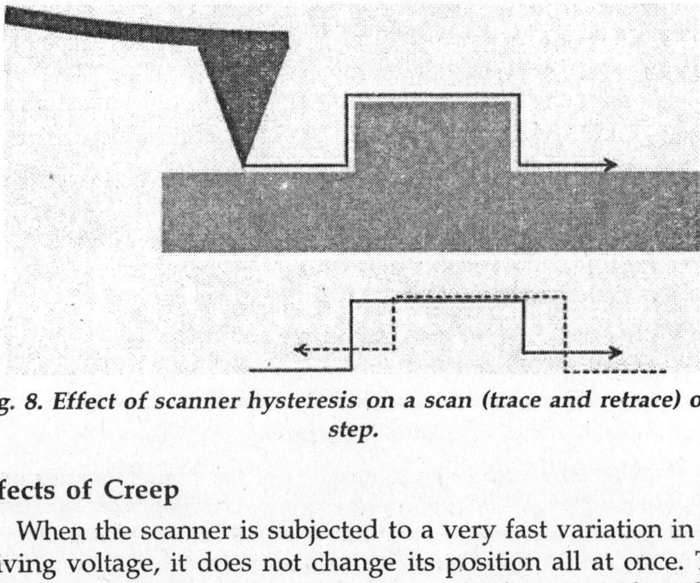

Fig. 8. Effect of scanner hysteresis on a scan (trace and retrace) of a step.

Effects of Creep

When the scanner is subjected to a very fast variation in the driving voltage, it does not change its position all at once. The dimensional change occurs in two steps: the first step takes place in less than a millisecond, the second on a much longer time scale. The second slower movement is called creep. This causes several effects. Scans taken at different scan rates will have slightly different magnification. If one tries to zoom-up onto a feature, making a smaller scan just after a larger scan, the feature will not be centered and may be distorted in the second image because of creep. On a structure made of parallel lines the effect will be a bending of the lines in the first portion of the scanned image (Fig. 9). This is often also called drift, but must not be confused with thermal drift, which is different.

In the vertical direction, creep becomes apparent as an overshoot of the scanner position at the leading and trailing edge of features that have steep sides (Fig. 10). This can be often found as a lateral "shading" of protruding features on flat substrates in top view topographical images.

Effects of Cross Coupling and Sample Tilting

Usually scanners are assembled in the AFM having a free end that is scanned (to which either the cantilever or the sample is attached) and the other end is attached to the microscope body. For this reason the motion of the scanner will follow an arc (spherical or parabolic depending on the type of scanner) and not a plane (Fig. 11). The affected images will show a bow, which is especially evident in large scans.

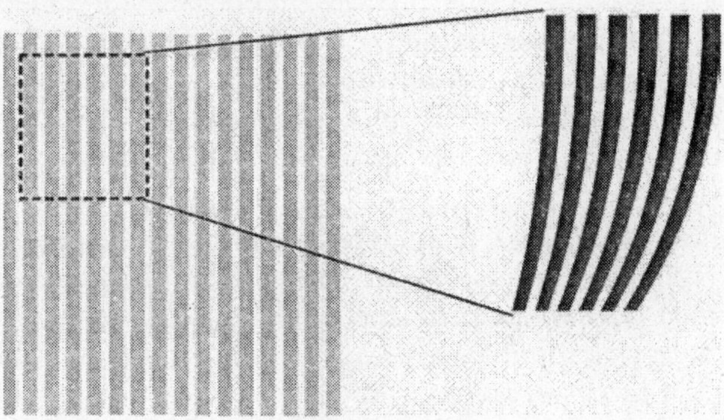

Fig. 9. Effect of creep on a scan performed zooming up onto a detail in a larger image.

This artifact can easily be subtracted by image processing. When very small features have to be detected on flat surfaces, the bow will not allow them to be seen during scanning as the vertical scale of the image would have to adjust to accommodate it: for this reason, the AFM often has the option to subtract the appropriate curve from each line during acquisition, allowing small features to become immediately evident. When there is mechanical or electronic cross coupling between the *x* and *y* direction elements of the scanner, this will become apparent in the image of test structures, where the angles between features in the *x* and *y* plane will be modified. Mechanical coupling between the piezoelectric ceramics that move the probe in the *x* or *y* directions and in the z direction can cause substantial errors when measuring sidewall angles. Another source of cross coupling arises when the scan direction is not parallel to one of the piezoelectric elements that constitute the scanner. Rotated scans are obtained by sending appropriately mixed driving signals to both the x and y piezoelectric elements: if they are not both accurately calibrated the image will be affected by a geometrical distortion.

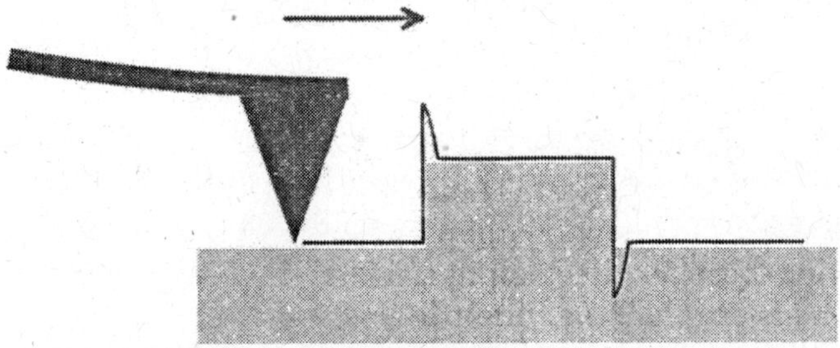

***Fig. 10.* Effect of creep in the vertical direction: overshooting at the edges of the step.**

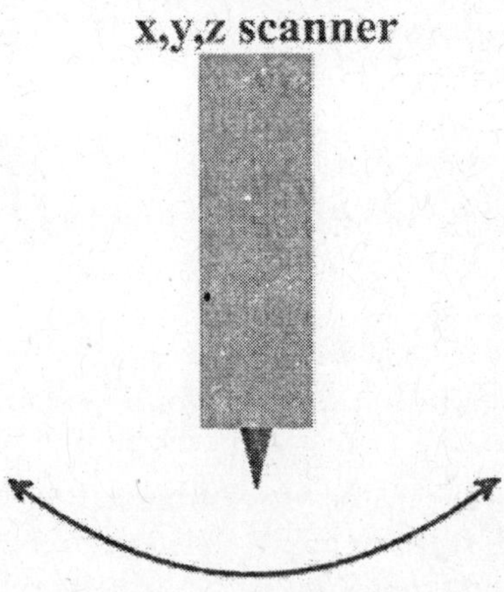

***Fig. 11.* The free end of the scanner will follow an arc during scanning, creating a bowl-like image. This effect is especially evident on large scans of flat surfaces.**

It is useful to add that quite often (in fact, always) the sample will have a plane tilt relative to the motion of the scanner. Although all acquisition software allows for subtracting the tilt during scanning, it is good practice to try and mount the sample as planar as possible so that the piezoelectric element responsible for the vertical movement will operate across a smaller range and hence behaving linearly.

Thermal Drift

External temperature changes or gradients will affect the AFM and its scanner depending on their mechanical properties. AFMs are built in such a way as to minimize this phenomenon by using special materials and appropriate design, but nevertheless thermal drift can be present. In the case of very high-resolution imaging of atomic structures, it is often necessary to wait some time during scanning before the system will stabilize and stop drifting. Also, electronics and the laser spot on the cantilever can induce drift in measurement settings that need to stabilize in time. In the case of AFMs mounted onto inverted microscopes and eventually equipped with a heated stage, special care has to be taken. When a liquid cell is used and reagents are flowed in during an experiment, the problem becomes acute as temperature changes of even a fraction of a degree can cause large bending of the cantilever. Often, a good ambient air conditioning system can be useful in reducing thermal effects.

VIBRATIONS

Because the AFM operates thanks to its very high sensitivity to the small deflections of the cantilever assembly, it is evident that if external vibrations affect the cantilever these will create artifacts in the images. Typically, the artifacts will appear as oscillations. Both acoustic and floor vibrations can excite vibrational modes in an AFM and cause artifacts.

The floor in a building can vibrate vertically several micrometers at frequencies below 5 Hz. The floor vibrations, if not properly filtered, can cause periodic structure in an image. This type of artifact is most often noticed when imaging very flat samples. Sometimes the vibrations can be started by an external event such as an elevator in motion, a train going by, or even people walking in a hallway. A special air table or bungee cords must be used to isolate the AFM from these vibrations. A good idea is also to install the instrument near a corner of the laboratory instead of at the center of a room, choosing if possible the lowest floor in the building.

A person speaking in the same room as the microscope, music, a door that shuts, an airplane going over the building can generate sound waves that will generate artifacts in the AFM images. Some instruments have as an option an acoustic hood or enclosure to isolate the AFM from external noise.

EFFECTS OF FEEDBACK AND OTHER PARAMETER SETTINGS

Depending on the mode of operation, several parameters have to be set by the user to obtain the best images. Among these, one can find deflection set point (in contact mode), oscillation amplitude and dampening (in AC modes), feedback gain (sometimes separated into a proportional gain setting and integral-derivative setting), low pass filters, scan speed, and so on.

The setting of these parameters is a trial-and-error process. Each time a new sample is put into the microscope, the best values must be searched and during the process many artifacts can be produced in images. Soft samples generally must be imaged at low scan speeds and low interaction forces, otherwise glitches in the scan direction or even sample deformation may occur. Rough samples again need to be imaged slowly, but larger amplitude or deflection might be needed to keep track of the surface. Especially in AC imaging modes (but also in DC mode) special care must be taken in tuning the gain parameters of the feedback. If the feedback loop of a scanning probe microscope is not optimized, the image can be affected. When feedback gains are too high, the system can oscillate, generating high-frequency periodic noise in the image. This may occur throughout the image or be localized to features with steep slopes. However, when feedback gains are too low, the tip cannot track the surface, and features will be distorted and smeared out. On large objects with sharp slopes, an overshoot can appear in the image as the tip travels up the slope, and an undershoot can appear as the tip travels down the slope. Taking a force-vs-distance curve to ascertain the presence of adhesion forces or other effects can help to guide the choice of imaging parameters.

IMAGE PROCESSING

Image processing is readily available in AFM as the data is stored digitally on a computer disk. One can easily access routines for flattening, polynomial-line or surface subtraction, removal of bad data, matrix filtering, and three-dimensional representation with sophisticated rendering. Often some kind of processing will be necessary to analyze data and compare it with other results, but care must be taken to avoid introducing artifacts. The most common ones stem from careless use of the powerful image processing tools available. For example, as we have seen in Subheading 3.4., nearly all images are affected by a tilt and by a bow introduced by the scanner geometry. If the wrong curve fit is applied or if large features are not excluded from the surface subtraction parameter computation (all image analysis software allow to include or exclude surface area portions from the computation), distortions will be introduced. This is particularly true with line-by-line curve fit and subtraction.

Low-pass filters, although capable of reducing noise in the data, will introduce smoothing of sharp features and, in the worst cases, delete smaller details. Fourier transform and power spectrum filtering if misused can create periodic features that may seem to be atomic structures, wheres in reality they are only noise.

SOME GUIDELINES FOR ARTIFACT TESTING

If during a measurement you get suspicious that an image may contain artifacts, here are some things you can do to be sure whether or not they are present:

- Take more that one image of the same area or the same line to ensure that it looks the same. When looking at a single scan line profile during acquisition, look if the traces are identical and stable in time.
- Try changing the scan direction and take a new image. You can do this also on a single scan line looking at the profile and observing directly the difference between the trace and retrace plots.
- Change the scan size and take an image to ensure that the features scale properly.
- Rotate the sample and take an image to identify artifacts induced by the shape of the tip.
- Change the scan speed and take another image (especially when suspicious periodic or quasiperiodic features are present). If they scale, you are looking at periodical noise.

4

Advanced Biosensing Using Micromechanical Cantilever Arrays

INTRODUCTION

During the last few years, a series of new detection methods in the field of biosensors have been developed. Biosensors are analytical devices that combine a biologically sensitive element with a physical or chemical transducer to selectively and quantitatively detect the presence of specific compounds in a given external environment.

These new biosensor devices allow sensitive, fast, and real-time measurements. The interaction of biomolecules with the biosensor interface can be investigated by transduction of the signal into a magnetic, an impedance, or a nanomechanical signal. In the field of nanomechanical transduc-tion, a promising area is the use of cantilever arrays for biomolecular recognition of nucleic acids and proteins. One of the advantages of the cantilever array detection is the possibility to detect interacting compounds without the need of introducing an optically detectable label on the binding partners. For biomolecule detection, the liquid phase is the preferred one but it has been shown that the cantilever array technique is also very appropriate for use as a sensor for stress, heat, and mass. Recent experiments showed that this technique could also be applied as an artificial nose for analyte vapors (e.g., flavors) in the gas phase.

NANOMECHANICAL CANTILEVER AS DETECTORS

The principle of detection is based on the functionalization of the complete cantilever surface with a layer that is sensitive to the compound to be investigated. The detection is feasible in different media (e.g., liquids or gas phase). The interaction of the analyte with the sensitive layer is transducted into a static deflection by inducing stress on one surface of the cantilever as the result of denser packing of the molecules or a frequency shift in case of dynamic detection mode as a result of changes in mass.

OVERVIEW OF THE TWO DETECTION MODES

Static Mode

In static mode detection, the deflection of the individual cantilever depends on the stress induced by the binding reaction of the specific compounds to the interface. The interface has to be activated in an asymmetrical manner, as shown in Fig. 1. Most often one of the cantilever surfaces is coated with a metallic layer (e.g., gold) by vacuum deposition techniques and subsequently activated by binding a receptor molecule directly via a thiol group to the interface (e.g., thiol-modified DNA oligonucleotides) or, as in case of protein recognition, by activating the fresh gold interface with a self-assembling bifunctional bioreactive alky-thiol molecule to which the protein moiety is covalently coupled.

The radius *R* of the curvature of the cantilever is given by Stoney's law:

$$<3 = £Y^{2}_{cant}[6/?(1-y)]^{-1} \qquad (1)$$

where a is the stress, y is the Poisson ratio, *E* Young's modulus, and t_{cant} the thickness of cantilever. The thickness of the lever is an important parameter that can be varied to increase or decrease the sensitivity of the device. By reducing the thickness of the cantilever a larger deflection is achieved. Reducing the thickness by factor '2' increases the bending signal due to stress at the interface by factor '4'. The interaction of the ligand with the receptor molecule has to occur inmediately on the interface. No flexible linking of the receptor molecule is allowed as a result of the fact that the induced stress will be diminished. The receptor molecules should be presented in a tightly packed manner on the interface to interact with the substances to be analyzed.

Dynamic Mode

In the case of dynamic mode detection, the resonance frequency of the individual cantilever, which has to be excited, depends on the mass. The binding reaction of the analyte to the interfaces increases the mass, and the resonance frequency is normally decreased. In Fig. 2, the scheme of dynamic cantilever detection is shown. The cantilever is excited by a piezo element. The change in mass (Am) during the experiment as the result of an uptake of interacting biomolecules induces a change in the resonance frequency of the cantilever, which can be described by the following formula:

$$Am = k(4nt^2)^{-1}(f_1^{-2} - f_0^{-2}) \qquad (2)$$

where the resonance frequency before and during the experiment are f_0 and f_1, k is the spring constant of cantilever, and n is a factor dependent of the geometry of the cantilever. The uptake of mass as a result of specifically interacting molecules is doubled in this manner, and the cantilever does not respond to temperature changes via a bimetallic effect. Additionally, the preparation involves fewer steps as in the case of the static detection mode.

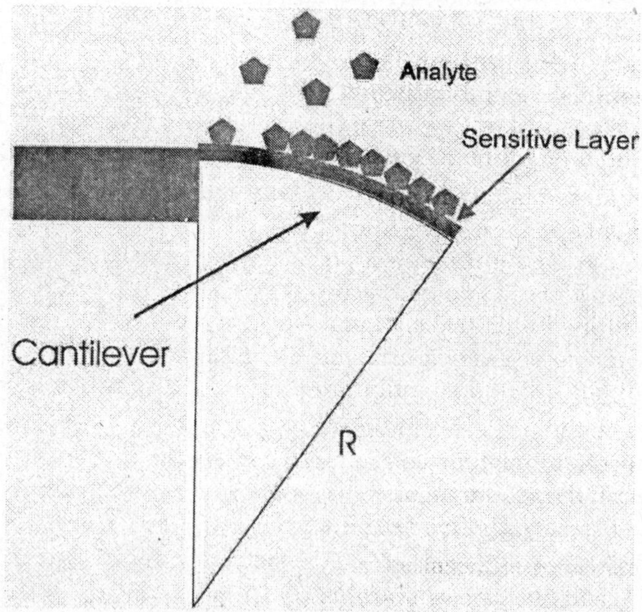

Fig. 1. Interaction of the analyte (light gray pentagons) with the sensitive layer induces a stress on the interface and bends the cantilever (note the asymmetric coating of the individual cantilever surface).

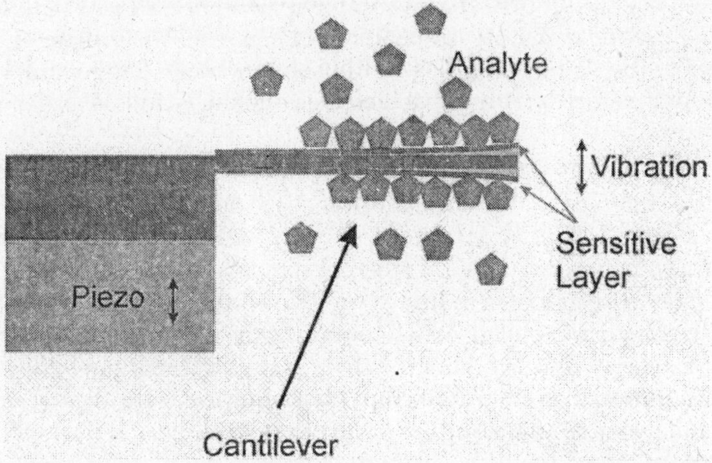

Fig. 2. Interaction of analyte (light gray pentagons) with sensitive layers induces a change in the resonance frequency of the cantilever.

SETUPS

At the Institute of Physics at the University of Basel, Basel, Switzerland, in collaboration with the IBM Research Laboratory Zurich, we developed cantilever array setups both for static and dynamic mode operation in liquids and in the gas phase.

The principal part of the setup is an array of eight cantilevers produced by classic lithography technology with wet etching. A typical picture of such a cantilever array is shown in Fig. 3. The structure of an array is composed of eight cantilevers with a length of 500 l im, a width of 100 l im, and a pitch of 250 l im from lever to lever. The etching process provides cantilever thickness ranging from 250 nm to 7 l im adapted for the individual application (i.e., static or dynamic mode).

The cantilever deflection or motion detection is provided by a classic laser beam deflection optical detection for both the static and dynamic mode set up as shown in Fig. 4.

The laser source is an array of eight vertical-cavity surface-emitting lasers (VCESLs; 760 nm wavelength, 250 l im pitch), and position detection obtained through a linear position-sensitive detector). The array is mounted in a cell useable for gas or liquid phase measurement. A scheme showing the setup is displayed in Fig. 5. The operation of the instrument is fully automatic and during the time course of a few hours up to eight different samples can be probed using the automatic fluid delivery. The instrumental noise of the static setup lies in the subnanometer range and the dynamic setup is able to detect mass changes in the order of picograms.

The key advantages of cantilever arrays are the possibility of *in situ* reference and the simultaneous detection of different substances. The *in situ* reference is needed to avoid the thermomechanical noise, especially in fluid-phase detection. Changes in refractive index when the buffer changes will also contribute to a so-called virtual motion of the cantilever. As visible in Fig. 6, only the real motion, which is the difference in between the cantilevers on the same chip, is originating from the specific biomolecular interaction.

In Fig. 7A, a raw signal of the cantilever array is displayed. Because there will always be instrumental or thermal drift, the differential signal detection is mandatory. Figure 7 shows an experiment with a set of three cantilevers (thickness 500 nm).

Fig. 3. Scanning electron micrograph of an array of eight cantilevers with individual thicknesses of 500 nm.

In this experiment we used two reference cantilevers with different coatings and one specific biorecognition cantilever. By specifically binding biomolecules the cantilever is bending downwards due to stress generated on its surface. As visible in Fig. 7B, the differential signal lacks any external influences except for the specific biomolecular interaction, which induces a differential signal of approx 90 nm relative to the *in situ* reference. The experiment is reversible and can be repeated using different concentrations of analytes. In a recent work we presented data that allow the extraction of the thermodynamics of the interacting biomolecules (i.e., DNA). Deflection signals as small as a few nanometers are easily detected. Currently, the detection limit in static experiments lies in the range of nanomolar concentrations but can be significantly lowered in the future by using cantilever arrays in the range of 250–500 nm of thickness.

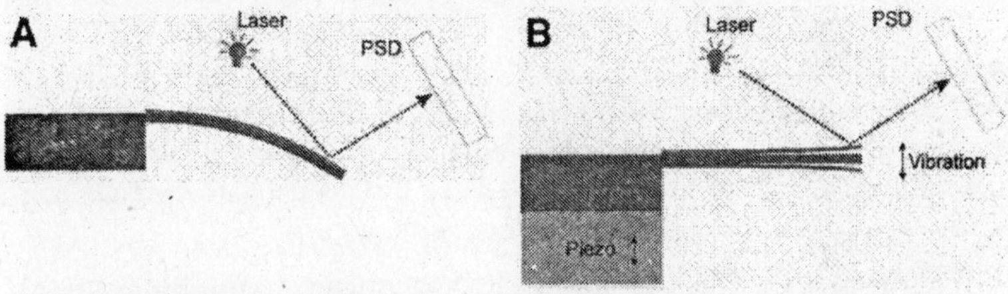

Fig. 4. Detection of average cantilever position using a multiple laser source vertical-cavity surface-emitting laser and a position-sensitive device. (A) Static mode; (B) dynamic mode.

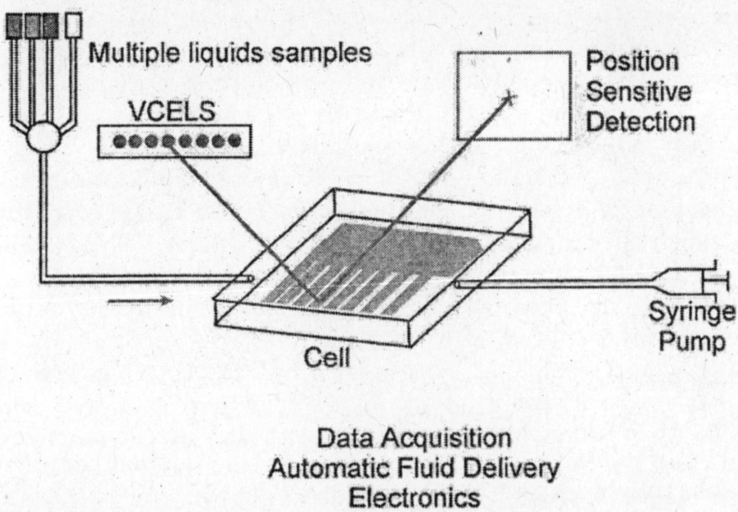

Fig. 5. General structure of cantilever array setups for gas/liquid samples.

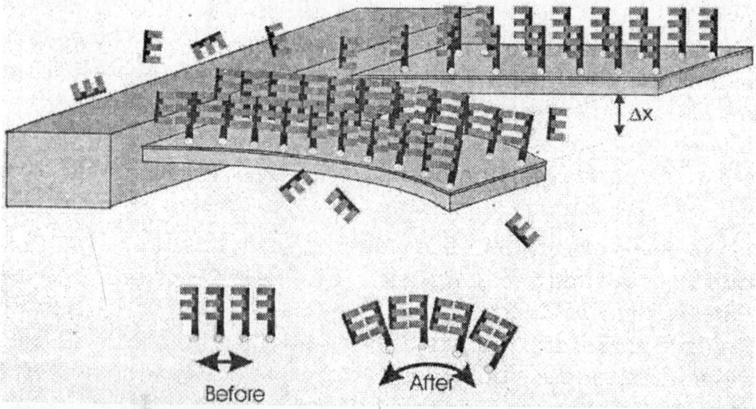

Fig. 6. Static detection of biomolecular interaction. The cantilevers have to be equilibrated before the biomolecule of interest is injected. Because of the specific interaction with the biomolecules (light gray) on the cantilever shown in front, stress builds up that deflects the individual cantilever specifically.

Great care has to be taken in the selection of the internal reference lever. In the case of DNA detection, an oligonucleotide is chosen that displayed a sequence that does not induce crosstalk binding reactions with the sequences to be detected. Coating with thin layers of titanium and gold using vacuum deposition modifies one side of the cantilever array. Onto this metallic interface, a thiol-modified oligonucleotide self-assembles in a high-density layer. Complementary and unknown oligonucleotide sequences are then injected and the specific interaction is directly visible within minutes. Stress at the interface is built up because of a higher density of packing. In protein detection, a protection of the asymmetrically coated cantilever has to be considered. Preparation of protein-detecting cantilevers is a multistep procedure and requires surface chemistry knowledge. The side opposite to the biomolecular-modified side is generally protected by a polyethyleneglycol layer. The bioreference surface can be coated by using unspecifically interacting proteins (e.g., bovine serum albumin). In protein detection experiments, larger fluctuations of the cantilevers are observed than in the ssDNA–ssDNA experiments. A possible interpretation of this difference might be that it is caused by the proteins absorbing light within the visible spectrum and therefore inducing some local changes in the index of refraction. We always measure specific signals within minutes without problems. Normally, some drift of few tens of nanometers is observed in the complete set of cantilevers during the time course of the experiment, even though temperatures of the instruments are stabilized within ±0.05°C. However, these effects are completely eliminated by using a differential read out on the very same cantilever array.

Cantilever arrays are already employed as detectors in both static and dynamic modes. Recent articles show the potential for detection of DNA hybridization, cell capture, or toxin detection. Integrating cantilever arrays into microfluidic channels will significantly reduce the amount of sample required. Attempts have been made to get data from single-cantilever experiments for DNA or antibody–antigen reactions or from a two-cantilever setups using different stiffnesses for the individual cantilevers. We would like to point out that these approaches have serious drawbacks. Information extracted from these experiments, which often last multiple hours, cannot exclude unspecific drift of any kind.

The signal in these experiments is interpreted as specificity on the biomolecular level but no correlation from one lever to the next is applicable if only one lever is used at a time. In the second approach cantilevers with different stiffnesses are used to monitor the nanometer motions. Because the individual cantilever used shows a difference of factor four in terms of stiffness.

CONCLUSION

The cantilever array technology explores a wide area of applications; all biomolecular interactions are in principle able to be experimentally detected using cantilever array as long as mass change or surface stress is induced by the specific interaction. A few applications so far demonstrate promising results in the field of biological detection. The cantilever-based sensor platform might fill the gap between the sensitive but costly and relatively slow analytical instrumentation (e.g., mass-spectroscopy, high-performance liquid chromatog-raphy, surface plasmon resonance [SPR]) and the chip technologies (for example, gene-arrays) with their advantage of easy multiplexing capabilities, albeit with their need for fluorescence labeling and restriction to higher molecular-weight compounds like proteins and nucleic acids thus far.

In comparison with the methods just described, the cantilever technology is cheap, fast, sensitive, and applicable to a broad range of compounds. The lack of multiplexing could be overcome by the application of large cantilever arrays with >1000 cantilevers per chip. Projects are now underway to introduce commercial platforms providing arrays of eight cantilevers to applications in the liquid or gas phases. A critical point for future development in this field will be the access to the cantilevers arrays similar to that in the normal field biological applications using single-cantilever scanning force microscopy. At the moment there are no biological experiments published that use dynamic mode detection. We believe, however, that its ease of preparation (symmetrically as pointed out) and reduced sensitivity to environmental changes, makes this technology a strong candidate as the instrumental approach of choice for the future biological detection using cantilever arrays.

5

Analysis of Human Fibroblasts by Atomic Force Microscopy

INTRODUCTION

The force-sensing members of the large family of scanning probe microscopies have become important tools during the past decade for visualizing, characterizing, and manipulating objects and processes on the meso- and nanoscale level. The atomic force microscope (AFM), in particular, has had an impact in the life sciences. In cell science, the pioneering work with AFM was conducted in the early 1990s. The methodologies have now reached a stage of relative maturity. The principal merit of the AFM is as a nonintrusive local probe of live cells and their dynamics in the biofluid environment. As well as offering high spatial resolution imaging in one or more operational modes, the AFM can deliver characterization of mechanical properties and local chemistry through operation in the force-vs-distance (F-d) mode. The lateral resolution delivered by the AFM will in most cases, and especially for soft materials, be inferior to that obtained by electron-optical techniques, but the *z*-resolution is routinely in the nanometer range with a depth of focus equal to the dynamic range of the *z*-stage travel.

The instrument may be operated in one of several modes, of which the most common ones are as follows: the contact mode, using a soft lever in which contours of constant strength of interaction are traced out; the intermittent-contact mode, in which a relatively stiff lever is vibrated at a frequency near that of a free-running resonance and in which contours of constant decrement of the free-running amplitude or a constant phase shift are mapped;

and the F-d mode, in which the local stiffness of interaction between tip and specimen is determined over a range of applied force (lever deflection and z-stage travel being the two measurable variables).

MATERIALS

Cell Culture: Handling and Preparation for In Vitro Analysis by AFM

Primary human skin fibroblasts and 3T3 cells are generally maintained as monolayer cultures in Dulbecco's modified Eagles' medium/Ham F12 containing 15 m*M* $NaHCO_3$, 50 U/mL penicillin, 50 mg/mL streptomycin, and supplemented with 10% inactivated (by heat) fetal calf serum (FCS). The cultures are then grown in a humidified environment with a 5% CO_2 atmosphere at 37°C. There are two reliable methods for preparation of live cells for AFM analysis.

Live Cells—Method 1

The cells are harvested by trypsin treatment in preparation for AFM analysis and are then grown overnight on sterile untreated glass cover slips. The cover slips are washed five times with serum free media (Dulbecco's modified Eagles' medium/Ham F12) containing 20 m*M* HEPES at pH 7.3. The cover slips can then be attached to standard AFM mounting plates before being covered with a droplet of fluid, in anticipation of in vitro analysis.

Live Cells—Method 2

The cells are grown overnight in sterile tissue culture dishes 150- or 40-mm diameter (e.g., Corning, cat. no. 430599). The preferred dishes have been pre-treated by the manufacturer so as to present a negatively charged and hydro-philic surface and thus ensure better cell adherence. The cells are then washed, as in the previous paragraph, and covered to a depth of 2–3 mm with serum-free media.

Fixed Cells

The preparation follows a modified procedure of Pietrasanta et al. The cells on cover slips are washed for 1 min in prewarmed (37°C) phosphate-buffered saline (PBS), and then for another 1 min in stabilizing S-buffer (0.1 *M* PIPES, pH 6.9, 0.5 m*M* $MgCl_2$, and 0.1 m*M* ethylenediamine tetraacetic acid). After stabilization the cover slips are washed again in PBS before fixing in 0.01% gluteraldehyde (in S buffer) for 10 min. This is followed by another wash in PBS before being thoroughly rinsed in deionized and distilled water. Finally, the specimens are dehydrated with ethanol (70, 80, 90, and 100% in sequence for 5 min per step). The specimens are then allowed to dry in air, whereupon they are ready for AFM analysis.

Dehydrated Cells

The preparatory sequence is identical to that described in the introductory paragraph. However, instead of placing the cultures in a biocompatible liquid for analysis, they are now rinsed to remove excess salts, allowed to dry in air for 30 min or more, and then investigated by AFM over periods up to 3 h. In some cases, it may be possible to continue analysis for periods up to 48 h.

AFM Instrumentation and Methodology

1. Instrumentation. The examples of AFM analyses described in Subheading 3. were performed with a ThermoMicroscope TMX-2000 multitechnique scanning probe microscopy system. Some of the results were obtained with a Discoverer Stage attachment using an open fluid cell defined by a droplet of water trapped between the glass window on the detector stage and the cover slip substrate (similar to that described in ref. 3. In this configuration, a 70 × 70 μm^2 scanner with a z range of some 13 μm was used. Any one of the mainstream AFM instruments will offer a comparable facility. On other occasions, an Explorer Stage attachment provided increased flexibility and convenience. Again, an open cell could be defined by a trapped droplet as in the case of the Discoverer Stage. It is preferable to grow the cell culture directly in a culture dish that will then constitute the fluid cell for the Explorer Stage. Several of the AFM manufacturers now offer comparably equipped standalone instruments or equivalent optional attachments. The preferred scanner for the latter procedure should have a maximum field of view of some 130 × 130 μm^2 and a z range of 10 μm. All combinations of instrumentation will

accommodate the full range of relevant operational modes: contact, *in situ* intermittent-contact in fluid, and F-d analysis.

2. Probe. A range of probes, available from several suppliers, can be used for contact-mode imaging and F-d analysis. Live cells, especially the plasma membrane, are soft objects in the context of AFM analysis (with an effective Young's modulus of less than 1 MPa). Accordingly it is necessary to work with a lever with a spring constant, k_N, in the range 0.001–0.1 N/m to avoid excessive tip indentation. The lower end of that range is suitable for general imaging and F-d analysis of the unsupported plasma membrane, whereas the upper end may be used for imaging the harder elements of the intracellular and cytoskeletal structures. Typical choices include (1) a V-shaped Si_3N_4 lever, with integral square pyramidal tip (nominal radius of curvature, R_{tip} = 40 nm, and aspect ratio, A_r. = 0.7), and with nominal values for k_N of 0.03 and 0.06 N/m; (2) a V-shaped lever, with integral triangular pyramidal tips with smaller tip radii (R_{tip} = 10–20 nm), and greater aspect ratio (A_r = 3), and with k_N = 0.5 N/m. The actual spring constant for a particular lever can be calibrated in anticipation of F-d analysis in accord with a method described in the literature *(7)*. Other methods will offer similar information *(8)*.

METHODS

Fluid Ambient Environment

Maintenance of Static Conditions

The principal variables requiring control are temperature and fluid volume. If the fluid cell is defined by a trapped droplet, then frequent replenishment of the reservoir is required. Thus, imaging conditions will need to be reestablished at regular intervals (typically 30 min), but there is then opportunity also to reestablish optimum temperature.

A larger fluid cell, such as a culture dish, with a volume of some 5 mL or more, will have a longer life span with respect to evaporative losses, and the greater thermal inertia will promote temperature stability. However, long-term stability over some hours will require replenishment. The optimum temperature can then be reestablished by total replacement of the media. Although flow-through replacement from an external reservoir is another option, the imaging conditions are likely then to be affected when a soft lever is being used.

Another alternative is that of continuous heating of the cell by a hot stage, the disadvantages then being associated with thermal contraction/expansion and with thermal convection currents in the fluid.

Injection of Reactants for Dynamic Studies

When slow dynamics are being investigated, reactants may be introduced when there is replenishment or replacement of media. A practiced operator can usually reestablish imaging/analysis conditions within a few minutes.

More rapid injection and mixing is required if fast dynamics are being investigated. Access to the fluid cell is generally restricted by the compact design of most instruments. Accordingly, a flow-through cell arrangement may be the better choice when interruption of the imaging conditions cannot be tolerated or when a particular field of view needs to be tracked continuously.

Fixed or Dehydrated Cells

When a cell is fixed, through cross-linking of the plasma membrane and/or dehydration, it becomes a hard object. Consequently, it can be imaged in air by routine AFM procedures; a typical contact mode image of a fixed cell is shown in Fig. 1. However, scanning electron microscopy (SEM)/tunnelling electron microscopy (TEM) are richer sources of information and are generally to be preferred for this type of investigation.

Live Cells In Vitro

Live cells on untreated cover slips rarely produce acceptable contact mode images because of poor adherence to the substrate, gross fouling of the tip, and destructive tip–membrane interactions.

However, cells grown on surface-treated culture dishes offer greatly improved imaging conditions. Continuous scanning at linear scan speeds of 150 µm/s and at force loads in the low-nN range can now be conducted over several hours without any apparent damage to viable cells. A typical example of a contact mode image of a lamellipodial region of a fibroblast is shown in Fig. 2; the contour line reveals intracellular structure with a *z* resolution in the low-nm range.

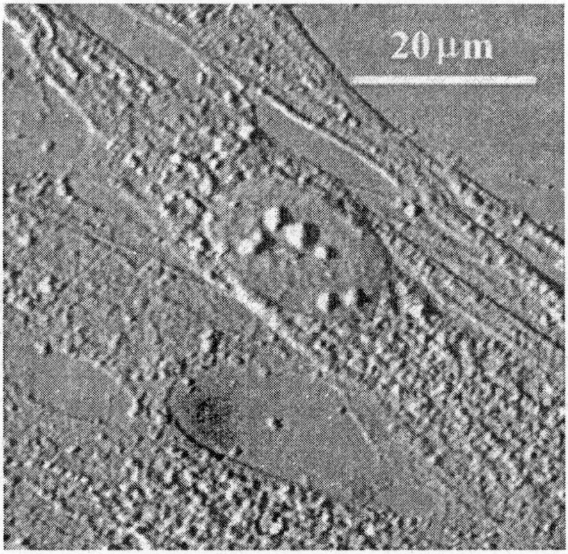

***Fig. 1** Example of a fixed fibroblast imaged in air in the contact mode.*

Intermittent-Contact Resonance Mode Imaging

It is argued that the intermittent-contact resonance mode (also known as tapping mode) will result in better image quality for soft specimens. The mode has two distinct advantages. The effect of lateral forces is effectively eliminated (important because cells have low resistance to shear stress). Because the interaction now arises from an impulse action at one extreme of the oscillatory motion of the tip, the inertia of the sample will resist deformation. However, the resolution of resonance mode imaging depends on the stiffness of the lever (i.e., the free-running frequency) and the width of the resonance envelope.

The latter is severely degraded in water, and soft levers are preferred for analysis of live cells. Although resonance mode imaging has many advantages, the jury is probably still out in the case of in vitro analysis of cells *(9–11)*.

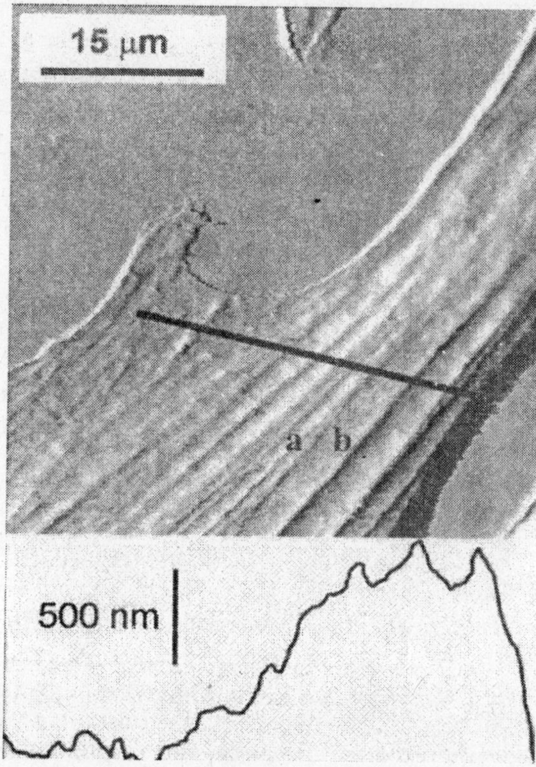

***Fig. 2** Lamellipodial region of a live human fibroblast in its biocompatible fluid medium on a culture dish substrate. Contact mode imaging was performed with a standard probe having a pyramidal tip (radius of curvature of 40 nm) and a lever spring constant k_N = 0.03 N/m. The lever-imposed force loading was approx 3 nN and the fast-scan speed was 150 µm/s. The contour line reveals submembrane structure.*

Analysis of Cell Dynamics

As mentioned in Subheading 1., the AFM is uniquely capable of tracking the temporal evolution of systems consisting of living cells in a biocompatible fluid. There are distinctly different methodologies for slow (>10 min) and fast (<10 min) dynamics.

Slow Dynamics

A skilled operator can establish good imaging conditions within a few minutes. The acquisition of an image takes typically

1–3 min. Thus, sequential imaging over a particular field of view can track cell dynamics in vitro on the time scale of some minutes. As in the cases described in Subheading 3.1.2., a soft lever (k_N < 0.01 N/m) in combination with a low applied force (<1 nN) will enhance information arising from the softer elements of the cell, whereas a stiffer lever and greater applied force will deform the plasma membrane and enhance visualization of the less compressible cytoskeletal and intracellular structures. The more informative studies have exploited the latter strategy to gain insight into cytoskeletal dynamics (e.g., refs. *5,12–14*). An example of investigations of slow intracellular dynamics is shown in a sequence of images in Fig. 3, where the intracellular nucleation and growth over a period of 3 h of formazan crystals is apparent. The crystals arise from enzymatic conversion of a tetrazolium salt during the MTT (3-[4,5-dimethyl thiazol-2-yl]-2,5-diphenyl tetrazolium bromide) assay of viable cells.

Fast Dynamics

Biological activity on the sub-second time scale can be observed and analyzed by AFM methodologies by monitoring the deflection of a lever stationary in the *x–y* plane and sensed effectively in the constant height mode (i.e., where the time-constant of the feedback loop is longer than that of the biological response mechanism). The tip is simply landed at an appropriate location predetermined from an image. The *x–y* scan function is deactivated, and the dynamic response is monitored through the *z* deflection of the lever. A soft lever is most appropriate because the probe is ideally a passive participant in the temporal evolution. The method has been deployed with considerable success in the case of cardiomyocytes *(15,16)*. There is clearly considerable scope for applications of similar methodologies for investigations of other manifestations of cellular dynamics.

ANALYSIS OF MECHANICAL PROPERTIES

F-D Measurements

The interaction between a tip and a surface is a function of the force imposed by the lever and manifests itself as a force acting on the tip while being sensed by the lever. Such a system consists of two compliant elements–the lever and the surface–whereas all other components of the system are assumed to be rigid. Thus, compressional forces will cause bending of the lever and indentation of the surface by the tip. If the shape of the tip, the spring constant of the lever, the applied force, and the depth of indentation are known, then an effective Young's modulus for the surface can be calculated.

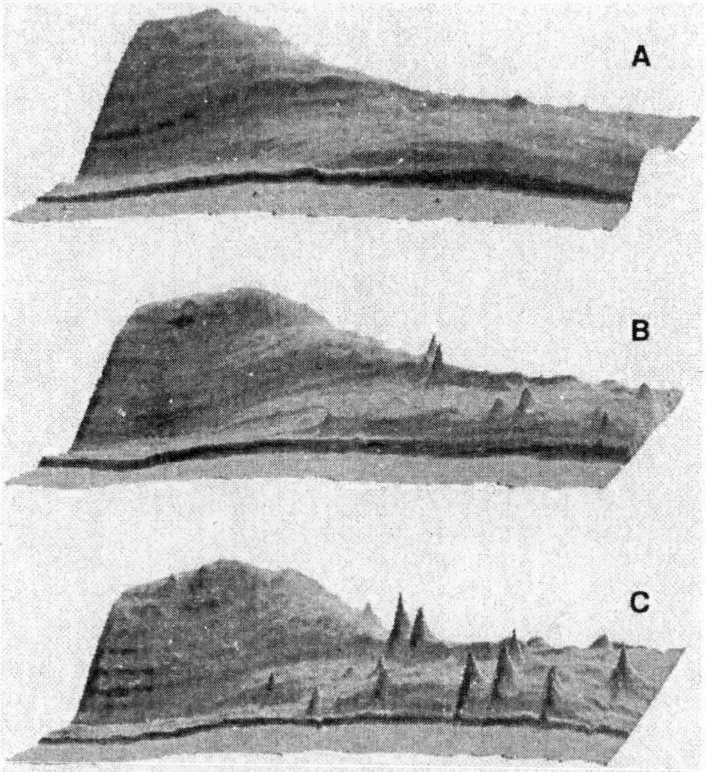

Fig. 3. Sequence of contact mode images from a lamellipodial region obtained in vitro showing nucleation and growth of formazan crystals within a viable cell. Images acquired after (A) 30 min; (B) 90 min; (C) 180 min.

Characteristic features of an F-d curve in combination with measurements of *z* stage travel and lever deflection allow such calculations to be conducted. The details of the procedure have been described elsewhere (e.g., refs. *5,17,18*). Adhesion between tip and surface manifests itself as an attractive force, causing deflection of the lever at the point of lift-off in the F-d curve.

Because the strength of adhesion is a reflection of the local surface chemistry, then mapping of the surface adhesion of living cells may provide valuable additional information.

F-D Methodology for Fibroblasts

The effective Young's modulus of a supported section of the plasma membrane is in the range 1–10 kPa, whereas the corresponding value for a membrane more strongly supported by the cytoskeletal structure is in the range 15–50 kPa. The effective modulus of fixed cells is an order of magnitude higher. To obtain reliable information about mechanical properties, it is necessary that the spring constant of the lever be comparable with the effective force constant of interaction between tip and specimen. Thus, k_N in the range 0.01 N/m must be chosen. Given the extreme softness of the plasma membrane there is no incentive for working with sharp tips; even lever-imposed forces in the sub-nN range will give rise to tip indentations of 2–40 nm and contact areas of greater than 100 nm^2. Adhesive interactions add to the lever-imposed force and cause additional indentation and yet greater contact area.

Generic Outcomes of F-d Analysis of Live Fibroblasts

The results in Fig. 4 illustrate typical outcomes of F-d analysis in vitro of fibroblasts. A standard probe with k_N = 0.03 N/m and a pyramidal tip shape, nominal radius of curvature of 40 nm, and an aspect ratio of 0.7 was used for the measurements. The data illustrate tip indentation, differences in stiffness measured at different locations, a and b, on the plasma membrane, the relative incompressibility of a fixed cell, and adhesive interactions between tip and membrane. The horizontal axis refers to z-stage travel, whereas the vertical axis shows lever deflection in the *z* direction.

Summary of AFM Studies Conducted on Fibroblasts

Table 1 summarizes critical methodological aspects of earlier AFM-based studies of flbroblasts and provides references for additional information.

Notes

1. Cell adhesion to substrate. Cell adhesion remains a significant methodological issue, although treated tissue

culture substrates will generally offer acceptable performance. The extent of adhesion can readily be ascertained by running single line scans repetitively over a representative part of a cell. In extreme circumstances, a line scan will remove the entire cell from its substrate. The reproduc-ibility of features in successive line scans is a useful monitor of stability of the cell during actual imaging conditions, as well as providing a measure of nondestructiveness of tip–cell interactions. Optimum quality of the image can also be obtained through adjustments of fast-scan speed and direction, force loading, parameters of the feedback loop, and field of view.

2. Tip contamination: effects and diagnostics. A cell cultured in a biofluid contains proteins, cell debris, and other contaminants in solution. The probe tip will inevitably become contaminated, at the very least, by nonspecific adsorption of proteins. Biofouling of the tip will alter the surface chemistry of the tip, and thus potentially its a destructive adherence to the cell, as well as its topography, with consequential degradation in resolution. The latter is not a serious problem, in the case of biomolecular adsorption, since extreme lateral resolution is not

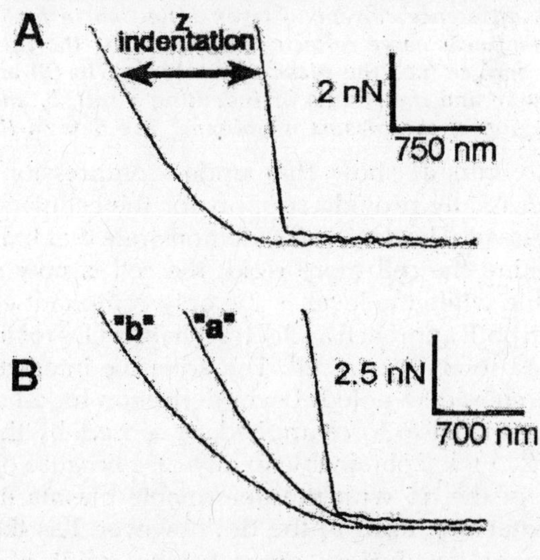

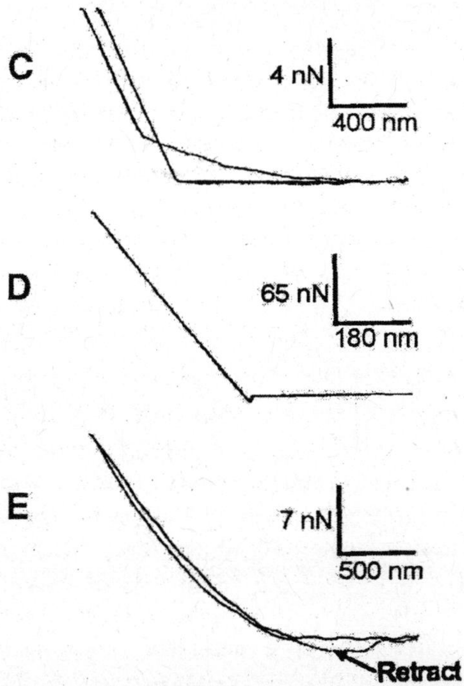

Fig. 4. (A) F-d curves obtained in vitro for a live fibroblast for different conditions. The steeper curve shows deflection of the lever by an incompressible surface and was used to calibrate the detector (relating nA difference current to lever deflection in nm). The less steep nonlinear curve reflects indentation by the tip in a compressible surface (i.e., the plasma membrane). In (B) are shown a calibration curve and the results of indenting a soft, b, and a stiffer, a, region of the plasma membrane. The data in (C)

Fig. 4. *(continued)* show that under compression the hard substrate will finally provide support for the cellular structures being compressed. The data in (D) demonstrate that fixing has the effect of making the cell more rigid; the cell is now effectively incompressible while the lever is the only compliant element. In (E) are shown both approach and retract half-cycles for a biofouled tip being used to analyze a cell. The adhesive interaction in the retract curve should be noted. Lever deflection for all curves can readily be converted to force applied, or sensed, by the lever by multiplying Z_L by k_N. obtainable in any case because of the large contact area of the tip with the deformable plasma membrane. Cell debris attaching itself to the tip, however, has the effect of reducing image resolution, often to the point of complete obliteration. Here are several ways of diagnosing tip fouling, aside from its effect on the image quality. Because the general topography of the substrate can be determined at any time with a fresh tip, any subsequent deterioration in definition of topographical resolution must be caused by tip fouling. A more quantitative method is to conduct reverse imaging of the tip *(8,26)*, whereby an image of the tip is generated from a scan over a spiky feature (e.g., an upturned tip attached to a substrate). Figure 5 shows reverse images of an as-received tip, and of a tip after exposure to a biofluid. Finally, a contaminated tip may be analyzed in the F-d mode by indenting on a known hard substrate. If the tip is compliant, as a result of adherent biodebris, then it will be obvious from the F-d curves. 3. Common image artifacts. Several of the early studies have reported prominent effects because of precipitation of salts from the biofluid solution. If the analysis is conducted in an open cell, and the cell is subject to evaporative losses, then the solution will become supersaturated in salts. Consequently, crystalline precipitates will form within the field of view. Moreover, the biofluid will no longer be compatible with cell viability. Frequent replacement of the biofluid will substantially eliminate that problem.

Table 1: Methodological Aspects of Past Studies of Fibroblasts

Cell type	Substrate	Subst. coating	AFM	Probe(s)	Re
Fibroblasts, etc.	Glass cover	None	Contact	Si3N4k_N=0.0 1,0.03	*20*
L929	Culture dish	None	Conta ct,	Si3N4k_N= 0.03,0.06	*21*
Fibroblasts,	Plastic	None	Contact	Si3N4k_N = 0.03	*13*
3T3	Plastic dish	None	Conta ct,	Si3N4k_N = 0.08	*22*
3T6	Glass cover	None	Conta ct,	Si3N4k_N = 0.01	*23*
3T3, NRK	Plastic dish	None	Conta ct,	Si3N4k_N = 0.08	*14*
NIH3T3	Glass dish	Fibronect in	Conta ct,	Si3N4k_N = 0.018	*24*
NIH3T3	Glass dish	Fibronect in	Contact, F-	Si3N4k_N = 0.03	*25*

[a]A viscoelastic modulation method was adopted for the particular study.

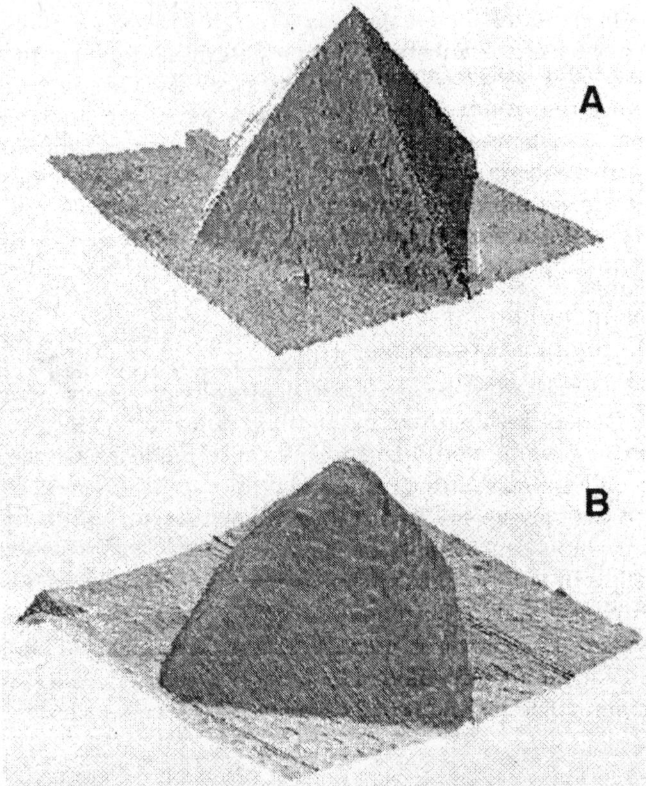

Fig. 5. Reverse images of a probe as-received (A) and after exposure to a biofluid (B).

Tip-broadening and other tip-related artifacts will occur when the actual topography of the object being imaged is defined by radii of curvature less than or comparable to the radius of curvature of the tip, and/or when there are gradients exceeding that corresponding to the aspect ratio of the tip. For instance, images of tobacco mosaic virus (TMV) attached to a flat substrate obtained by AFM reveal the correct height of approx 18 nm, but the apparent lateral width will be in the range 60–100 nm as a result of the tip-shape convolution *(27)*. Because the radius of the cylindrical TMV is known and is comparable to that of the apex of the tip, the apparent width of the object, *W*, in is given by the following:

$$W = 2[(R_{TMV} + R_{Tip})^2 - (R_{Tip} - R_{TMV})^2]^{1/2}$$

When cytoskeletal structure is being imaged, the situation is somewhat more complicated by filamentary objects located some distance above the substrate. The aspect ratio then comes into play because the deformable membrane allows the tip to indent the cell on either side of the filamentary object. The apparent width will now depend on the height, *h*, of the object above the point of greatest indentation by the tip on either side of the object. The relevant expression is now as follows:

W H" $2[hA_r^{-1} + (r_{tip} + r_{obj})\cos$ ö]

where the radii of the tip and object are r_{tip} and r_{obj}, respectively; A_r, is the aspect ratio of the tip, and the angle is defined by ö = $\tan^{-1} A_r^{-1}$.

Finally, other grosser artifacts will occur when the dynamic range of the *z* stage is exceeded; the image then becomes entirely featureless. A similar effect occurs when the *z*-height corrugations of the object exceed the height of the tip, and the surface of the lever defines the point of contact. The interaction is no longer localized, and the details of the image become washed out. Likewise, F-d analysis will now produce erroneous data since the spring constant will depend on an unknown and changing point of contact and the contact area will also be much greater leading to erroneous conclusions about indentation and adhesion.

6

Corneal Tissue Observed by Atomic Force Microscopy

INTRODUCTION

The cornea is the transparent avascular part of the anterior segment of the eye and consists of a stratified nonkeratinizing squamous epithelium, a stromal dense connective tissue layer, and an endothelium facing the anterior chamber. The cornea contributes largely to the intraocular refraction of the light. Damage can impair its tissue transparency and lead to loss of vision. Significant diseases, such as corneal dystrophies, keratoconus, and refractive errors, are related to the structure and integrity of the cornea.

In conventional scanning electron microscopy studies, the corneal surface appears like a mosaic consisting of three types of cells, as it can be deduced from their electron reflex and size *(1–4)*. The apical membrane of these cells is covered by the tear film. The inner corneal surface, facing the anterior chamber of the eye, is the apical membrane of the endothelium, which forms a monolayer of polygonal cells responsible for maintaining the state of relative deturgescence of the stroma through active transport *(5–10)*. The stromal layer consists of regularly arranged dense connective tissue constituting 90% of the corneal thickness. It comprises sheets of lamellae of highly ordered collagen fibrils, embedded in a matrix of proteoglycans, and keratocytes. The former are interspersed between the lamellae, forming an interlinking network throughout the cornea *(11–13)*.

AFM has been recently introduced with success in the research of corneal surfaces and components *(11,14–16)*. Compared with other forms of microscopy used in corneal study, AFM offers several advantages: it can reach very high magnifications with high resolution, it requires minimal tissue preparation, and it is able to image samples in aqueous environments, thus permitting images to be obtained under conditions that resemble the tissue's native environment. Additional advantages include the possibility of dynamic in vivo study of biological processes and the capability of characterizing the nanomechanical properties of relatively smooth surfaces. Limitations of the method include the relatively small scan sizes and scan speeds and difficulties in imaging very soft biological samples. Because of such limitations, the AFM is currently used either as an investigational tool or as an adjuvant to other microscopic techniques. In long term, however, it has the potential to evolve in a unique multipotential instrument for the study of the morphology and mechanical properties of various biological tissues.

This chapter describes the methodology used to study the surface of the cornea in albino New Zealand rabbits and in humans. We describe the procedures necessary in rabbits to study the normal epithelial and endothelial surfaces as well as the corneal stroma after mechanical and excimer laser ablation. Samples were imaged in balanced salt solution (BSS) both fresh and after fixation in glutaraldehyde. We studied in humans the endothelial surface of two corneal buttons received after corneal transplantation for endothelial dystrophy. The tissue was imaged in BSS after fixation in glutaraldehyde.

MATERIALS

Tissue Collection and Preparation

1. Rabbit corneas: New Zealand albino rabbits with 3–4 kg body weight.
2. Human corneas: Transplant recipient corneal button.
3. Anesthesia solution: 10 mg/kg xylazine hydrochloride + 10 mg/kg ketamine hydrochloride.
4. Proparacaine drops.
5. Operating microscope.
6. Surgical blades.
7. Excimer laser.

8. Surgical instruments for enucleation and corneal dissection.
9. Precision wipe paper.
10. Rinsing and observation solution (Alcon Laboratories, Fort Worth, TX).
11. Solutions for enzymatic preparation: 30 mU/mL neuraminidase in phosphate buffer solution (Sigma Chemical Co., St. Louis, MO); 30 mU/mL hyaluronidase in phosphate buffer solution (Sigma).
12. Fixative solution: glutaraldehyde, 2.5% buffered solution, pH 7.3, at 4°C.
13. Buffer solution for fixative preparation: 0.2 *M* stock solution of sodium cacody-late, pH 7.3, kept at 4°C.
14. Euthanasia solution: sodium pentobarbital.

Microscopy Equipment

1. AFM (Nanoscope IIIa, Digital Instruments, Veeco Inst., Santa Barbara, CA), including an optical viewing system and image analysis software.
2. Piezo-electric scanners, 12–150 μm.
3. V-shaped silicon nitride tips with a spring constant of 10 mN/m (Microlever; Park Scientific Instruments, Sunnyval, CA).
4. Magnetic stainless-steel punches.
5. Epoxy glue.
6. Fine forceps for tissue transfer and manipulation.

METHODS

Tissue Collection

Anesthesia

The animals are anesthetized with a subcutaneous injection of xylazine and ketamine. Additional topical anesthesia with proparacaine drops is used to anesthetize the cornea.

Stromal Ablation

The anesthetized animal is placed under the operating microscope. Mechanical ablation is performed using a sharp surgical blade, and the anterior one third of the cornea is dissected taking care not to penetrate the cornea. Excimer laser ablation is performed following a standard protocol for myopia correction; a myopic correction of three diopters is aimed.

Euthanasia

Animals are euthanized by an injection of sodium pentobarbital overdose delivered via a peripheral ear vein.

Enucleation

The eye globes are carefully enucleated as soon as possible after death. Special carè is taken not to contaminate the corneal surface with blood and not to touch or stress the tissue during manipulation. Eyes that will be imaged fresh are placed in BSS solution. For eyes that are going to be examined fixed, the fixation process described in the next paragraph is followed.

Human Corneas

The recipient corneal buttons from patients undergoing corneal transplantation are collected.

Fixation Process

Rabbit Eyes

Immediately after enucleation, the eye globes are placed into fixative solution. After 30 min and while the eye globe is still in the solution, a hole is opened 6 mm behind the limbus to allow penetration of the fixative solution in the interior of the eye. The fixative solution is replaced with freshly prepared solution. The eyes are kept overnight in the solution at 4°C before AFM observation.

Human Corneal Buttons

Immediately after trephination, the recipient button is placed into fixative solution. The eyes are kept overnight in the solution at 4°C before AFM observation.

Preparation of Corneal Specimens

Handle all cornea specimens withfine instruments under microscopic observation, paying attention not to distort the tissue during manipulations such as cutting, transportation, and gluing

Rabbit Corneas

This step is performed immediately after enucleation in eyes that are going to be imaged fresh. Fixed eyes are processed after completion of the fixation. The anterior part of the eye is cut away and the cornea is freed from the underlying iris, cilliary body, and lens. The tissue is trimmed near the sclerocorneal limbus and it is dissected in two semicircular pieces.

Corneal specimens are transferred to magnetic stainless-steel punches and are fixed with epoxy glue. Specimens are maintained with the surface that is going to be examined upwards. Before transfer, the excess of solution is absorbed from the seating side by using a precision wipe paper. After transfer to the magnetic punches all specimens are covered with BSS solution and placed under the micoscope. For corneas that will be observed after enzymatic treatment the process described below is followed prior to transfer to the punches.

Human Corneas

The corneal button is dissected in two semicircular pieces. Corneal specimens are transferred to magnetic stainless-steel punches and are fixed with epoxy glue. Specimens are maintained with the surface that is going to be examined upwards. After transfer to the magnetic punches, all specimens are covered with BSS solution and placed under the microscope.

Enzymatic Preparation

The cornea freed from the underlying iris, cilliary body, and lens is immersed in neuraminidase or hyaluronidase enzymatic solution with the surface to be examined directed upwards. The dishes containing the enzymatic solutions are closed and kept at 37°C for 30 min. After the completion of this time, they are removed from the solution and rinsed gently with BSS for 5 min to remove the excess of enzyme and the enzymatic digestion products. After that the specimens are transferred to magnetic punches.

AFM Imaging

Image Aquisition

1. The area of interest is chosen using the optical microscope attached to the viewing window of the AFM. The central

area at a distance of some millimeters from the specimen's edges is considered the area most appropriate for observation.

2. Imaging starts using large scanning areas, when possible. Large scanning areas provide information about the general topography of the sample and allow for the selection of flat regions without defects for small-scale imaging. For imaging of areas from 20–100 μm (Fig. 1) a 100-μm scanner is used. For smaller areas ranging from 10–0.2 μm, high resolution can be achieved with a 12-mm scanner (Fig. 2).
3. To obtain good images, the force curve needs to be corrected repeatedly. In fresh tissue the adhesion of the surface glycocalyx sugars to the microscope tip, results in fuzzy images. In these sample it is often difficult to achieve a good forces-vs-distance curve and several tries are necessary until satisfactory images are acquired (Fig. 3A). Imaging of fixed tissue is considerably easier because the surface glycocalyx is removed during the fixation process (Fig. 3B).
4. The scan rate ranges between 0.5 and 10 Hz, depending on the scan size. Small frequencies are used to scan large areas (Fig. 4) and vice versa.
5. Imaging forces of not more than 100 pN are used. High forces are applied only as a means to mechanically remove the surface layer that adheres to the tip.
6. Images are obtained with a resolution 512 × 512 pixels of trace and retrace collecting data. Three types of images can be obtained during the contact mode imaging:

a. In height images the color-coded contrast refers to the spatial variation of the Z-height of the tip.
b. In deflection images the contrast differences of the surface refer to the spatial variation of the strength of the probe–specimen interaction.
c. In lateral force microscopy or friction images, information concerning the friction on the surface f the specimen during the movement of the tip is displaced. However, interpretation and analysis of the later images of the cornea remains difficult.

Image Analysis

1. For a better presentation, height images are processed using a plane-fit adjustment, when the sample surface is not perpendicular to the scanner's z-axis.
2. To evaluate the surface structure, sections on the height images are used that present the profile of the surface. These sections are indispensable when features like protrusions, particles, holes, fibrils, and so on have to be measured. The sections are performed on raw data images. Zooming is necessary when small features of large-scanning images have to be measured.

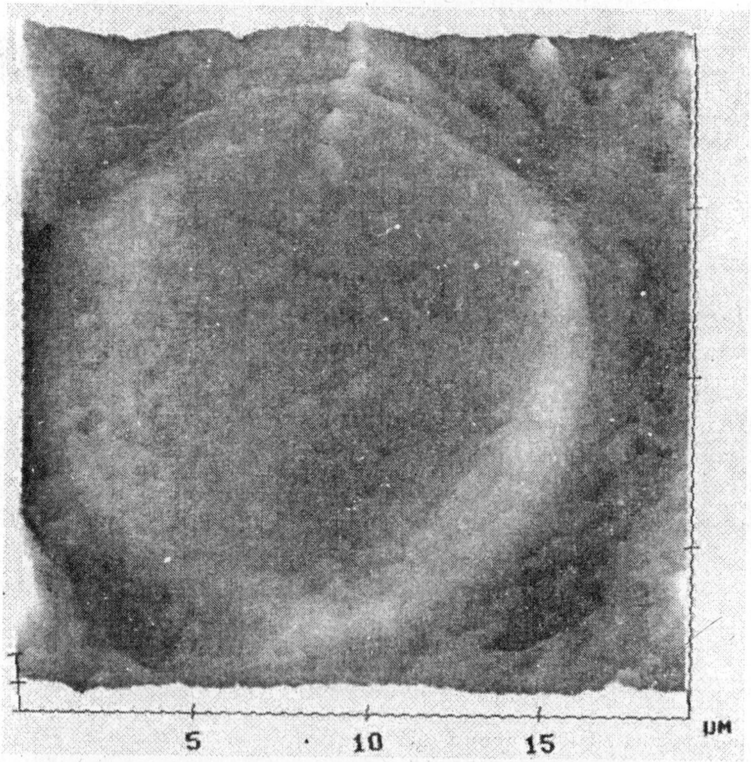

***Fig. 1.** Low-force contact-mode AFM image. Human corneal endothelium from a patient with corneal endothelial dystrophy who underwent corneal transplantation. The recipient corneal button was studied with AFM. 20-µm scan range; 1.5-Hz scan rate; scanning force <100 pN.*

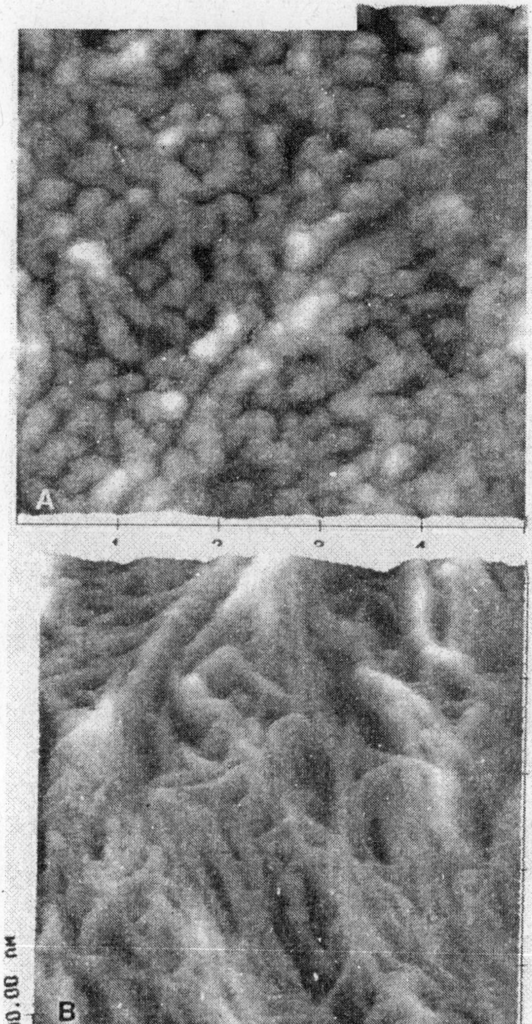

Fig. 2. **(opposite)** *Low-force contact-mode AFM images. (A) Height image of fixed rabbit corneal endothelium showing a detail of the intercellular contact of two epithelial cells and the micro-projections on their surface. 5-µm scan range; 2-Hz scan rate; scanning force <100 pN. (B) Height image of fixed rabbit corneal stroma after mechanical dissection. Collagen fibrils appear randomly arranged. In some of them the periodicity is apparent. 10-µm scan range; 2-Hz scan rate; scanning force <100 pN.*

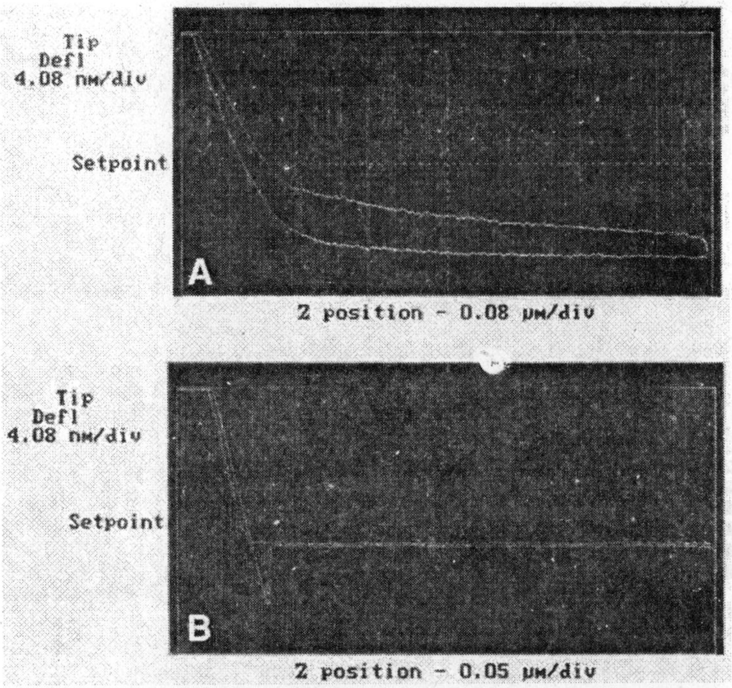

Fig. 3. (A) Force-vs-distance curve recorded on the fresh corneal surfaces. The cantilever's deflection in the vertical axis is converted into force using the relationship $F = k_{cl}y$ where k_{cl} is the spring constant of the free cantilever. (B) Force-vs-distance curve recorded on the fixed corneal surfaces. Note the difference between fixed and fresh specimen curves.

3. Quantitative data are acquired after the measurement of several morphological characteristics. The meta-analysis tools provided by the system's software facilitate for the calculation of statistical and topographic parameters. These include the ratio of the length along the longer axis over the height of measured structures as well as the measurement of surface roughness. Such quantitative analysis gives more precise information about the morphology of the surface.
4. Roughness statistics are performed on height images 5 × 5 µm. Mean roughness (*Ra*) and root mean square (RMS), or *R(q)*, are calculated. *R(q)* is the standard deviation of the Z values in a given area whereas*Ra* is the mean roughness value of the surface relative to the center plane.

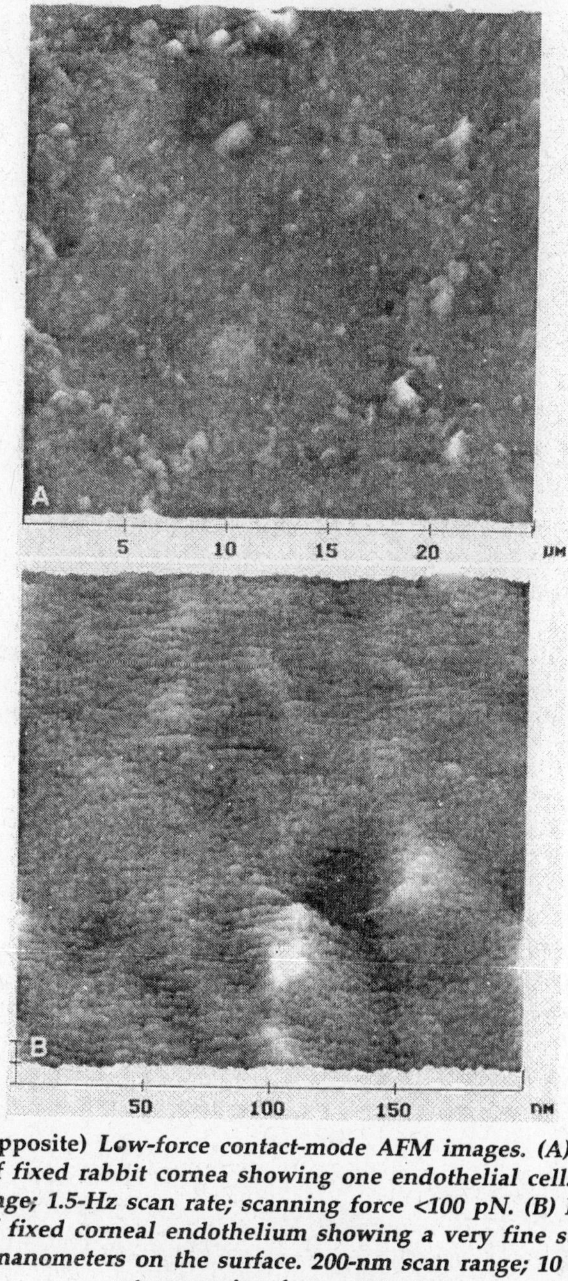

Fig. 4. (opposite) *Low-force contact-mode AFM images. (A) Height image of fixed rabbit cornea showing one endothelial cell. 25-µm scan range; 1.5-Hz scan rate; scanning force <100 pN. (B) Height image of fixed corneal endothelium showing a very fine structure of a few nanometers on the surface. 200-nm scan range; 10 Hz-scan rate; scanning force <100 pN.*

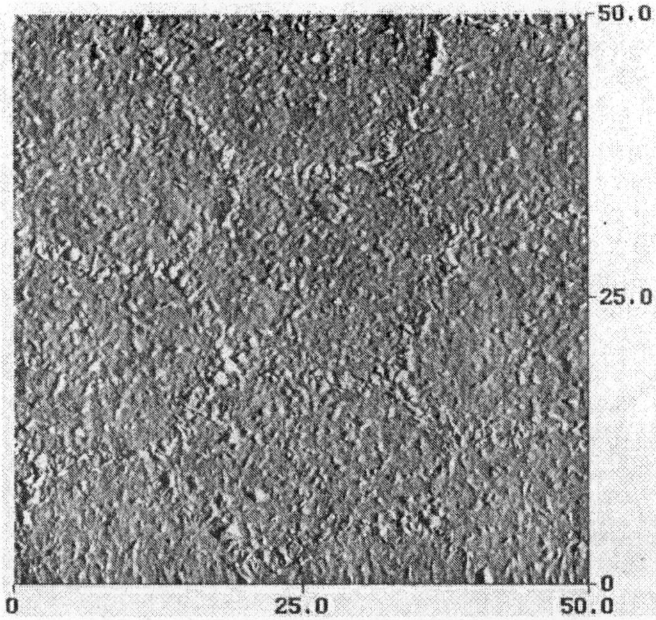

Fig. 5. Deflection AFM image. Fixed rabbit corneal endothelial surface. The contour of endothelial cells its easily detected. Because of their height contrast, these images are suitable for counting the features on the surface. 50-µm scan range; 0.5-Hz scan rate; scanning force <100 pN.

Notes: To be able to extract information from AFM imaging it is important to minimize the risk of artifacts before or during the imaging.

1. Before imaging, fresh tissue poses considerable imaging difficulties. The interaction of glycoaminoglycans chains of the glycocalyx layer with the microscope tip makes the imaging of the fresh tissue difficult (Fig. 6). When the imaging of a specimen is important not to fail, consider fixation. The removal of the glycocalyx that happens during fixation makes the imaging easier.
2. Enzymatic treatment represents another way to improve image acquisition from fresh tissue. The enzymatic process increases the unevenness of the sample surface thus increasing contrast. In addition, information concerning the sample's molecular composition can be revealed and

help in the interpretation of the effect of the enzyme on the surface morphology (Fig. 7).

3. When imaging of the outer corneal surface is intended, it is important to handle the tissue very carefully during preparation. Contamination and distortion of the superficial corneal layers can happen very easily and will alter the surface morphology. When the tissue is going to be imaged fixed, the installation of a few fixative drops on the corneal surface while the animal is in deep anesthesia just prior to euthanasia ensures preservation of the superficial corneal layers in the best possible condition.

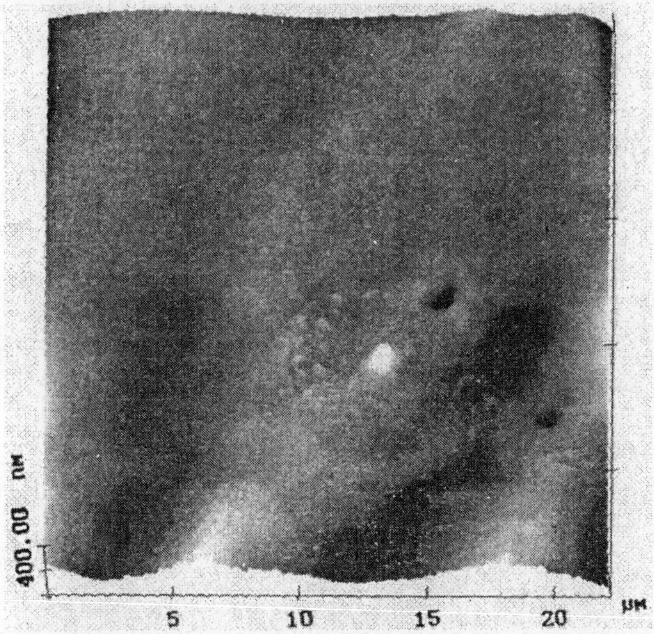

Fig. 6. ***Low-force contact-mode AFM image. Height image of fresh epithelium. 22-µm scan range; 1.5-Hz scan rate; scanning force <100 pN. A large part of the surface appears fuzzy.***

4. Gentle manipulation of the tissue in general is very important. Prepare all the instruments and materials in advance. It is essential to work under an operating microscope or a stereoscope especially when cutting the samples to be imaged. The working place, the instruments, and the solutions need to be very clean. Prior to imaging,

inspect the sample's surface and ensure it is not defective or contaminated.

5. Time optimization: tissue preparation, cutting, and gluing on the pounches must be completed as quickly as possible to avoid tissue drying.
6. BSS represents our preferred medium for observation. This solution was selected because it contains all the essential ions necessary for maintenance of the rabbit and human corneal integrity *(10,18,19)*.
7. Allow 15–30 min after the installation of the sample under the microscope for the system to reach a thermal equilibrium. This will elliminate thermal drifting.

Fig. 7. Low-force contact-mode AFM image. Fresh rabbit corneal endothelium treated with neuraminidase. A microgranular structure can be seen on the surface together with elevated aggregates of different sizes. 4-µm scan range; 3-Hz scan rate; scanning force <100 pN.

8. Adjust the level of the set-point force by using the force-vs-distance curve (Fig. 3). This determines the force that

the tip applies to the sample. A set–point level close to the jump-out point ensures an operation with minimal force.

9. When a soft cantilever is used, the applied force should be maintained in the sub-nano-newton level. Higher forces produce significant surface alterations. This effect is more pronounced in fresh tissue.
10. Duration of sample observation: in fixed tissue, the duration of observation of a specimen can be extended to 2.5 h without obvious morphological alterations. Observation of fresh tissue in BSS should not exceed 1–1.5 h. If the process is prolonged over this time the tissue hydration changes and edema occurs.
11. When there is any doubt concerning the tip's quality, check it and if necessary replace it.

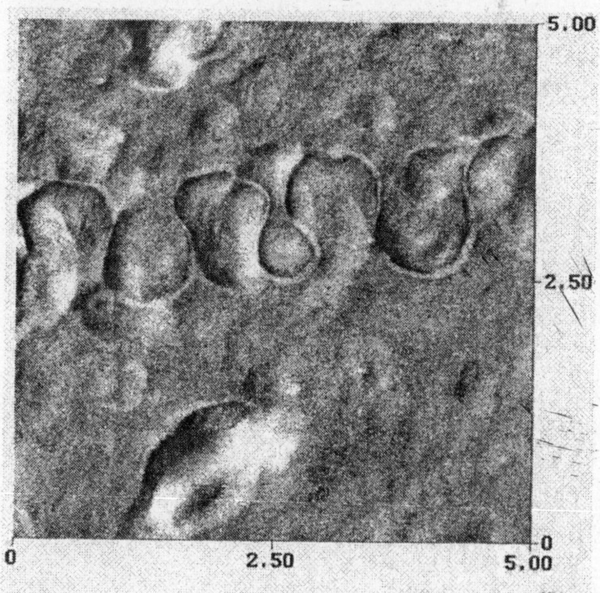

Fig. 8. ***Low-force tapping-mode AFM. Amplitude image of fixed rabbit cornea showing the corneal endothelial cell interdigitations. 5-µm scan range; scanning force <100 pN; 0.500-nm amplitude.***

12. When the specimen's surface is rough, it may be necessary to alternate various imaging conditions (scan angle, scan frequency, etc.) to obtain an image of acceptable quality.

13. If the surface does not permit imaging using a large scanning area it is preferable to reduce the scanning area. This way you will save time and you will reduce the risk of damaging the sample and/or the tip.
14. Zooming to a smaller scanning area should be accompanied by an increase in the scanning speed.
15. Imaging is usually performed in contact mode. Taping mode gives height-images with a quality inferior to that of contact mode. The phase and amplitude images in tapping mode are usually of very good quality (Fig. 8). For the time being, however, the interpretation of these images is not easy.

7

AFM Study of Surface Structure Changes in Mouse Spermatozoa Associated with Maturation

INTRODUCTION

If a sample has a comparatively even surface and is fixed on a sample stage, atomic force microscopy (AFM) will give a clear image of the surface structure at subnanometer level *(1,2)*. Because a sperm head is flat and can be attached on the slide glass firmly after it is fixed, we consider that AFM is the competent tool for the study of the sperm surface structure.

Spermatozoa are produced in the testis and transferred into the epididymis. In the epididymis, they acquire a fertilization ability and mobility, which is called sperm maturation *(3)*. It has been proved biochemically and immunocy-tochemically that glycoproteins on or in the sperm plasma membrane are altered, masked, or replaced by new glycoproteins of epididymal origin in the epididymal duct *(4–6)*. These changes are considered to be necessary for fertilization *(3,7)*.

Thus, it is probable that the sperm surface structure changes progressively in the epididymal duct *(8)*. We reported the changes in the surface structure of the spermatozoa in the hamster epididymis by AFM *(9)*. Subsequently, we have studied the surface structure of the spermatozoa from mouse epididymis by AFM. In these studies we developed several tricks for improving AFM images. In this chapter, we will show our recent results and the materials and methods used, including the tricks we learned in these studies.

Histologically, the mouse epididymis is divided into five regions (segments I–V) in adult male mice (Fig. 1; refs. *10,11*). These segments perform different roles in the process of sperm maturation *(12)*. Spermatozoa are immature in segment I but mature in segment V *(13)*. The epithelial cells in segment II appear to secrete glycoproteins for sperm maturation *(10,14,15)*, so we examined the segments I, II, and V of the mouse spermatozoa by AFM.

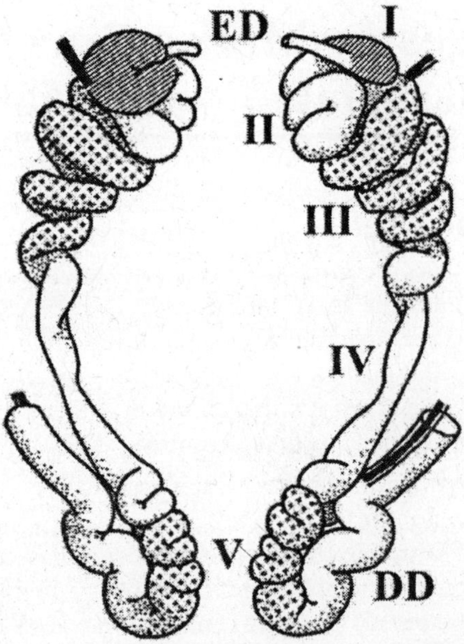

Fig. 1. Schematic diagram of the histological segmentation of the mouse epididy-mis. ED, efferent duct; I–V, epididymal segments; DD, deferent duct.

The surface of the sperm head is classically divided into two domains, the acrosomal region and postacrosomal region (Fig. 2). The acrosomal region is subdivided into the acrosomal cap and equatorial segment (Fig. 2). These domains play the different roles at fertilization *(16,17)*. The plasma membrane of the equatorial segment is maintained after acrosome reaction and fuses with the plasma membrane of an oocyte at fertilization *(3)*. Because of this, our study focused on the changes in the surface structure of the equatorial segment.

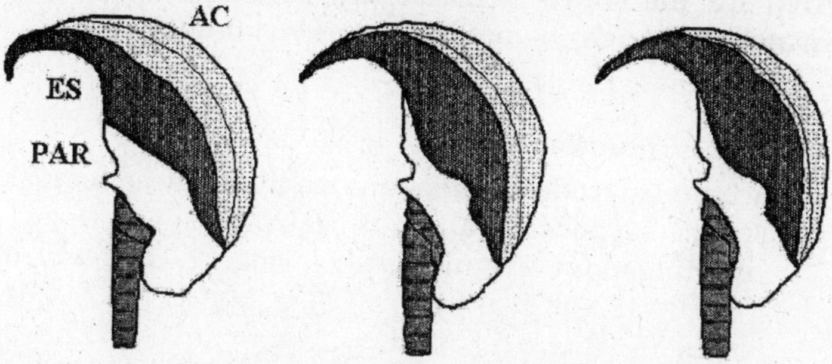

Fig. 2. ***Diagrams traced from the AFM images of the mouse spermatozoa from segments I* (left), *II* (middle), *and V* (right). *AC, acrosomal cap; ES, equatorial segment; PAR, post-acrosomal region.***

Changes in the Sperm Surface Structure

Dynamic force mode AFM, as used in our laboratory, provides images in constant height mode, amplitude mode, and phase mode. Images in constant height mode offer the topographical information of the sample height (Fig. 3). Those in amplitude mode and phase mode offer well-defined contours of the projections and hollows based on the changes in oscillation amplitude and phase, respectively (Figs. 4–6). Using a combination of these modes, the sperm head was shown to be covered with particles whose diameter was < 100 nm. The size of the surface particles differed between the acrosomal region and postacrosomal region in the same spermatozoon. Particularly, the particles covering the equatorial segment changed in size dramatically during the passage through the epididymal duct (Figs. 4–6). The surface of the equatorial segment was covered with the particles of 20–50 nm in diameter in segment I, 50–80 nm in segment II, and 20 nm in segment V. These size differences generated the different morphological surface features among the spermatozoa from segments I, II, and V (Figs. 4–6). Because the epithelial cells in segment II appear to secrete acid glycoproteins, which play a role as a sperm maturing factor *(10,14,15)*, large particles covering the segment II sperm surface may be gly-coproteins secreted from the epithelial cells in segment II.

AFM images also demonstrated the changes in shape of the acrosomal cap. The acrosome cap is flat and wider in the immature spermatozoa from segments I and II than in mature spermatozoa from segment V (Figs. 2 and 3A through 6A). Thus, an application of AFM for the study of the surface structure of the mouse spermatozoa brought us noteworthy new findings.

Materials

1. A mature male dd-mouse at 90 days of age.
2. Modified tyrode solution (this medium is used for making sperm suspension and for washing spermatozoa by centrifugation): 500 mL distilled water, 2.05 g NaCl, 0.1 g KCl, 0.1 g $CaCl_2$(anhyd.), 0.05 g $MgCl_2/H_2O$, 0.025 g NaH_2PO_4/H_2O, 1.5 g $NaHCO_3$, approx pH 8.0. This stock solution can be used for one month if stored at 4°C.
3. Fixative: 2% glutaraldehyde in 0.1 *M* cacodylate buffer.
4. Ethyl alcohol (anhyd.).
5. 3-methylbutyl acetate.
6. Nitrogen gas.
7. A double-edged razor.

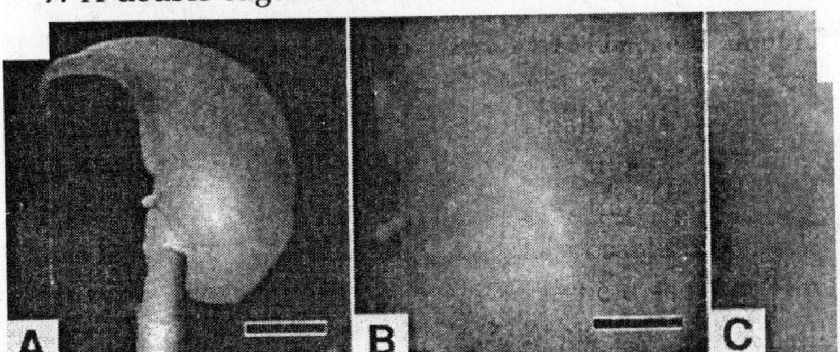

Fig. 3. AFM image in constant force mode of epididymal segment I.

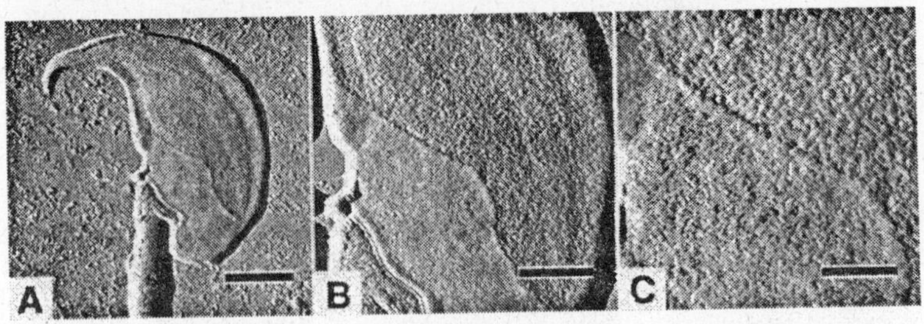

Fig. 4. AFM image in amplitude mode of the same spermatozoon as in Fig. 3.

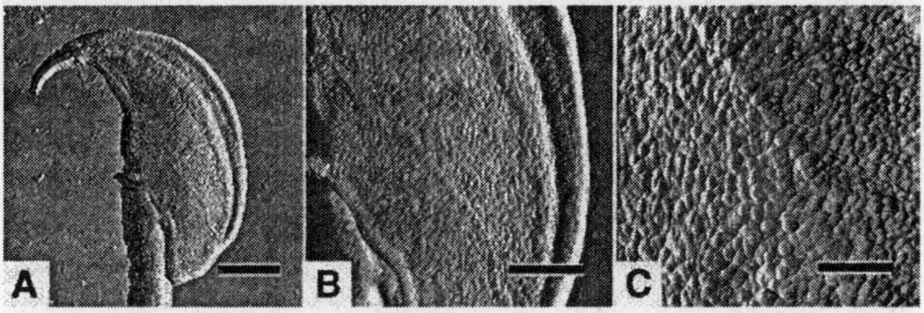

Fig. 5. AFM image in amplitude mode of the spermatozoon from segment II.

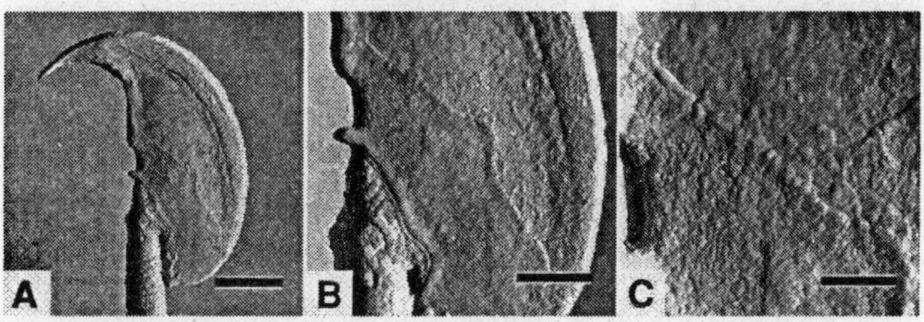

Fig. 6. AFM image in amplitude mode of the spermatozoon from segment V.

8. Small vials.
9. Test tubes.
10. Test-tube stand.
11. Disposable pipets.
12. Plastic dishes for ethyl alcohol.
13. A glass dish for 3-butyl-ethyl acetate.
14. Micro slide glass coated with adhesive material (Superfrost, Matunami Glass IND., LTD, Japan).
    a. Cut the micro slide glass into 7-mm squares.
    b. Mark the surface of one side of of the slide glass squares with the glass cutter to distinguish the side of the sperm attachment.
15. Centrifugal separator.
16. Critical point drier.
17. AFM.
    a. Probe station/Unit SPI3800 /SPA300 (Seiko Instruments, Japan).
    b. Scanner table (a maximum scan range is 20 μm × 20 μm × 3 μm, (x, y, z direction).

c. Lever table (20 N/m).
d. Cantilever tips SI-DF20 (Seiko Instruments, Japan).
e. Optical microscope.
f. Antivibration platform and nitrogen gas bomb.

METHODS

Preparation of Samples

1. Remove both sides of the epididymis from a mature dd-mouse. Separate the tissue of segments I–V of the epididymis (Fig. 1; ref. *10*).Cut tissue blocks of segments I, II, and V into small pieces with razors and place in the bottom of small vials separately.
2. Pour 3 mL of medium into each small vial and make the small pieces of tissue separate in the medium with tweezers. Wait a few minutes for spermatozoa to submerge from the stumps of the epididymal duct.
3. Transfer 2 mL of sperm suspension to a centrifugal tube and centrifuge at 240*g* for 8 min.
4. Pour 3 mL of medium into each centrifugal tube to wash the spermatozoa, and centrifuge at 240*g* for 8 min twice more.
5. Fix the spermatozoa in 2% glutaraldehyde in 0.1 *M* cacodylate buffer solution for 1 h.
6. Adjust the total amount of the sperm suspension to be 0.3 mL with distilled water after centrifugation.
7. Place one drop of sperm suspension on the slide glass square. As the glass is coated with adhesive material, spermatozoa adhere to the surface of the glass in 5–10 min. Make 3 or 4 samples.
8. Move the slide glass in distilled water to allow the release of loosely-attached spermatozoa from the slide glass.
9. Check the density of the spermatozoa on the slide glass under a light microscope.
10. Transfer the slide glass with spermatozoa in turn to 80 (5 min), 90 (5 min), 95 (5 min), and 100% (10 min, twice) ethyl alcohol solutions for dehydration. Figs. 3–6. *(facing page)* AFM images of mouse epididymal spermatozoa. (B) and

(C) are enlarged images of (A). Bar is 2 µm in (A), 1 µm in (B), and 0.5 µm in (C).

11. Immerse the samples in 3-butyl-ethyl acetate solution for 15 min.
12. Put a sample in a cage and perform critical point drying.

Scanning Operation Method of Dynamic Force Mode (DFM)

1. Allow the nitrogen gas to flow into the antivibration platform.
2. Turn on the antivibration platform and CCD camera controller.
3. Turn on the light of the optical microscope.
4. Set the scanner table.
5. Turn on the power at the center of the front panel of SPI3800N probe station controller.
6. Turn on the computer and the display. Microsoft Windows startup menu will appear.
7. Double click the icon of *Spise132* on the desktop to open the application for AFM.
8. Select "DFM" in "SPA300/400" frame and click "OK"; the main program, *SPIWin,* will start.
9. Set the sample stage on the mount on top of the scanner.
10. Set the sample on the sample stage.
11. Select a spermatozoon and move it to the center of the eye field of the optical microscope by hand.
12. Install the DFM cantilever holder with a DFM cantilever into the unit.
13. Select "CCD Monitor" in the setup menu to display "CCD". Check the position of the cantilever and the specimen by CCD image. Use the Move-in command (Low or High) and set the distance between the sample and the cantilever at approx 0.1 mm (which corresponds to a 180°- rotation of the fine-focus dial of the optical microscope).
14. Set the laser unit on SPA 300.
15. Adjust the laser light spot to be in right position.
    a. Rotate adjustment knobs "Laser X" and "Laser Y" to put the laser light spot on the top of the cantilever under the optical microscope.

b. Rotate adjustment knobs "Laser X" and "Laser Y" to maximize ADD output. The maximum value of ADD output is around 5 or 13 depending on the can tilever.
c. Rotate adjustment knob "DIF" so that the output may be in the range from –1 to 1 V
d. Rotate adjustment knobs "DIF" and "FFM" so that the output may be in the range from 0 to 1 V.

16. Measure the Q-curve.
    a. Select "Q-curve" in the Scan menu to display "Q-curve Console."
    b. Put the initial value of the parameter of the Q-curve Console. One example follows:

Freq. High	400 kHz
Low	1 kHz
Gain	1
Vib. Voltage	1 V
LPF	1 kHz
HPF	1 kHz
Time	5 s

    c. Select "Left" in the Vib. Freq. frame.
    d. Check "Phase," "Calibration," and "Auto Set."
    e. Click "Configuration" button to display configuration dialog.
    f. Set the value of the auto set. "Amplitude" is 1.000 V and "Frequency" is 3.000 kHz.
    g. Click "Start" to measure the Q-curve and the phase curve. Computer calcu lates the optimal value for the vibration frequency (operation point) immediately and displays the values with the Q-curve. An example follows:

Freq. High 128.410 kHz	
Low	125.410 kHz
Gain	1.000
Vib. Voltage 1.511 V	
LPF	1.0 kHz
HPF	1.0 kHz

Time	5 s
Vib. Freq.	126.800
Amplitude	1.068 V
Peak Freq.	126.918 kHz
AF	0.335
Q	379.273
Phase	-56.520°

h. Click "Close" to close, on the Q-curve Console

18. Approach the force area.
    a. Select "Image" in the Scan menu to display "Approach" and "Scan Console."
    b. Preset the parameters in the Scan Console follows:

       Amplitude Reference -0.101

       I-gain 0.177, P-gain 0.0488, A-gain 0, S-gain 0 when the maximum ADD value is around 5. I-gain 0.0592, P-gain 0.0122, A-gain 0, S-gain 0 when the maximum ADD value is around 13.
    c. Open "Sub Console" to confirm the filter values for topography.

       Data Type Topo (servo)

       LPF 1.000 kHz, HPF 0.000 Hz, Range 1583.20 nm, Offset 0.715V
    d. Approach the sample to the cantilever by using the Approach button.
    e. Lower the value of the Amplitude Reference until the scanning voltage becomes stable.
    f. Make the Amplitude Reference 0.3-0.5 down from the value of getting stability of the scanning voltage.
19. Separate the sample from the cantilever until the scanning voltage displays -20.
20. Turn off the laser light.
21. Insert the head of the spermatozoa under the tip of the cantilever under the optic microscope.
22. Perform the test scan at 512 x 128 points in the area of 15,000 nm^2 at the scan speed of 1 Hz.
23. Approach the force area by using the "Approach" button and click "Start."

24. Execute steps 21–23 again when the object is not displayed in the canvas.
25. Change the rotation angle, center of the image, and scan area to get the proper compositon. Use "Zoom" to change the center of the image.
26. Check the composition of the image by test scan.
27. Write the sample information in the column for comments.
28. Change the number of the scan points to 512 × 512.
29. The relation between "Scan Area" and "Scan Speed follows." *Scan area (nm) Scan speed (Hz)* 9,000

    0.28 8,000 0.28 6,000 0.37 4,000 0.41 3,000
    0.56 2,000 0.85 1,000 1.23

    It takes 40 min to get an image of 9,000 nm square.
30. Start to scan.
31. Set the S-gain value between 3 and 8 if the image is improved by this.
32. Separate the sample from the cantilever after completing the scan.
33. Save the measured data into HDD or MO as soon as possible.
34. When all measurements are finished, close the *SPIWin* software and *CCD* monitor.
35. Shut down the system.
36. Turn off the light of the optical microscope.
37. Turn off the power of CCD camera controller and antivibration platform.
38. Stop nitrogen gas to flow to the antivibration platform.
39. Turn off the power at the center of the front panel of SPI3800N Probe station controller.

Notes

1. When the ADD output does not move freely, readjust the position of the laser light on the cantilever or the distance between the specimen and the cantilever to be adequate (approx 0/1 mm).
2. The adequate value of the Amplitude Reference often moves in the minus direction during scanning. In this case

the scanned image will become blurred and disappear at last. To avoid this, the value of the Amplitude Reference is better set 0.3–0.5 lower than the value necessary to stabilize the scanning voltage.

3. Caution the following items to prevent the cantilever tip from damages.
   a. Scanning should not be stopped on the way of image delineation, even if the image is not good.
   b. The rotation angle and the scan area of the image should not be changed when the sample is in the force area.
   c. When the computer freezes, the sample should be separated from the cantilever by using up-down lever equipped in SPA300 before application is stopped.
   d. FFM knob should not be rotated during scanning.
4. Pay attention to the appearance of double images or two or more triangle-shaped particles arranged in the same direction, which reflect damage to the cantilever tip. If these signs appear, replace the cantilever immediately.
5. When an image does not appear in the NC-force canvas, try again and you will get the image.
6. A test scan is recommended after the scan area is changed because the scan position often shifts after that.
7. When the scanning voltage shows 200, the following three reasons should be taken into consideration.
   a. The laser beam is now off.
   b. The proper value of the amplitude reference is changed spontaneously to minus direction. In this case remeasurement of Q-curve is effective.
   c. FFM value is out of ±1.0 V.
8. Check the FFM value just before scanning, because FFM value is easy to change.

8

Calculation of Cuticle Step Heights from AFM Images of Outer Surfaces of Human Hair

INTRODUCTION

Atomic force microscopy (AFM) is an ideal technique for noninvasive examination of hair surfaces, providing a wealth of structural information not always apparent from electron microscopy. The fine cuticular structure of human head hair is of interest to those engaged in the fields of dermatology, cosmetics, and forensic science. In the former, the morphology of hair can be affected by an underlying inherited or congenital metabolic disorder, such as maple syrup urine disease or moni-lethrix, respectively. The cosmetics industry is interested in the effects of haircare formulations, such as conditioning and bleaching agents, on hair cuticle surfaces. There is now increasing legislation on cosmetic manufacturers to be able to substantiate claims made concerning their products.

Cuticle step height, as shown in Fig. 1, is an important parameter for the quantitative assessment of human hair. Step heights typically range from 300–500 nm but can vary further as a result of swelling or lifting caused by clinical, cosmetic, or environmental effects. This large variation in step height can be attributed to the heterogeneous character of hair cuticular structure.

The wide distribution of step height coupled with the need to obtain many step measurements clearly calls for a computational image processing technique. Such an approach is necessary to perform vast numbers of step height measurements for statistical comparisons. This becomes even more apparent when it is realized that there are a multitude of often-subtle differences in cuticle patterns between hairs from different parts of the head, between hairs from different body sites, and within each hair according to the distance from the skin surface. Except as a means for illustrating specific surface features, the single atomic force micrograph cannot be said to be representative of that hair and certainly not of a whole head of hair. This therefore focuses our attention on the need for quantitative extraction of surface architectural information, which by adequate sampling will enable systematic statistical comparisons to be made within and between hairs.

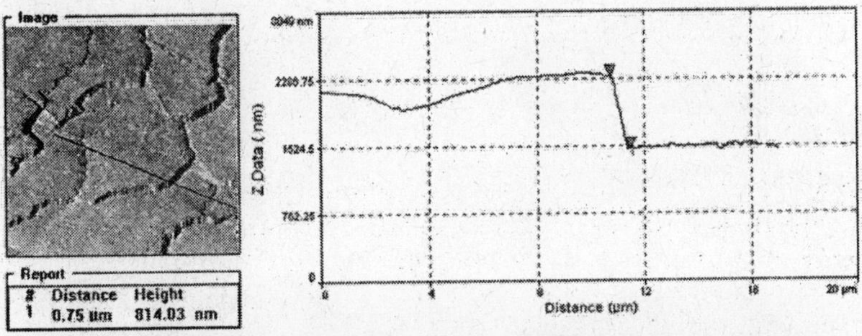

Fig. 1. A transect across an image of a hair cuticle showing a typical scale edge profile. In this case, the cuticle step height is shown to be 814 nm.

In this chapter, the method is demonstrated on a hair sample half of which has been bleached with a cosmetic formulation and the other left untreated. AFM topography imaging (Fig. 2) was conducted near the boarder region where the known length variation on surface architecture was considered to play a negligible role *(8)*.

Materials

1. A tress of human hair. In this chapter, a tress of European brown hair, of length 20 cm, was used.
2. Sodium dodecyl sulfate solution (1%).
3. Cosmetic formulation for hair treatment. Here, a commercial bleaching product was used.
4. Double-sided, adhesive carbon tape (*see* Note 1).
5. Tweezers for manipulating hair fibers.

6. Atomic force microscope. Here, a TopoMetrix TMX2000 Discoverer Scanning Probe Microscope, operated in contact mode, in air, using 200-μm V-shaped silicon nitride cantilevers (spring constant 0.032 N/m) was used.

METHODS

Sample Preparation

1. Place about 20 fibers in a 100-cm^3 beaker half filled with sodium dodecyl sulfate solution (1%), ensuring all the hairs are immersed.

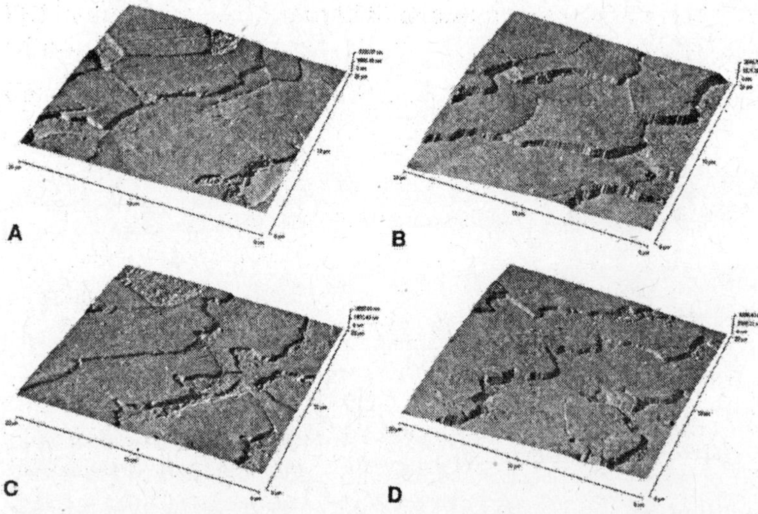

Fig. 2. Typical AFM topography images of European brown human hairs: two untreated hairs: (A) hair 1, and (B) hair 2; and two bleached hairs: (C) hair 1, and (D) hair 2. Examination of 10 images of each hair for both conditions suggests that the bleached hairs appear to be more damaged than the untreated hairs.

2. Place the beaker in an ultrasonic bath and sonicate at room temperature for 30 s.
3. Pour away the detergent solution, rinse the hairs with copious amounts of double-distilled water, and allow to air dry. Fibers can be subsequently stored between filter papers.
4. Hair samples should be fixed to an AFM mounting assembly (nickel stub) using double-sided carbon tape before AFM imaging.

AFM Imaging

1. The end of the cantilever should be placed over the center of the short axis of the hair specimen.
2. The scan range should ideally be limited to an upper value of 20 μm. This obviates the scanner from exceeding its *z* range while tracking the curvature of hair surfaces. Such artifacts have been reported elsewhere *(4)*.
3. The resolution should be set to 500, that is, the topography dataset will comprise 500 lines × 500 pixels. A greater resolution can be used if desired.
4. An optimal scan rate of 3 Hz is recommended.
5. Surface architecture information is best revealed using a hypothetical light source positioned to the left of the image.
6. Obtain an image of the hair surface and check the direction in which the cuticular sheets overlap one another.
7. Change the scan direction so that the cuticular sheets overlap from left (root end) to right.
8. Obtain 10 images each for treated and untreated regions for two hairs (4 × 10 images). Ideally, more hair fibers (approx 10) should be examined.

Image Analysis

1. Use the TopoMetrix Image analysis software (TopoMetrix SPM Lab. 1996, Version 3.06.06, TopoMetrix Corporation, Santa Clara, CA) to load the saved topography image and export the dataset as a text file. The images should not be levelled and/or shaded before exporting the data. The cuticle step height program requires the data to be delimited with comma separation and for the file header information (first 18 rows of data) to be retained. The 40 text files should be stored in a directory C:\subdirectory.
2. The text files should have three digit filenames. For example, "1b0.txt" refers to hair 1 (the prefix), bleached (b, or u for untreated), first image (0). Image numbers range from 0 to 9.
3. Use the cuticle step height program, coded in Microsoft MS-DOS QuickBasic, to calculate the cuticle step heights in all the images. A program listing is provided in Subheading 3.4. (*see* Notes 3 and 4).

4. Four output text files, consisting of one long column of cuticle step heights in nanometres (nm), are produced: hair 1, untreated; hair 1, treated (bleached); hair 2, untreated; and hair 2, treated.
5. The output text files can be read into a spreadsheet, such as Microsoft Excel, and imported into a software package such as Microcal Origin 4.2 to produce line graphs and histograms and to perform statistical analysis using, for example, oneway analysis of variance (Fig. 3).

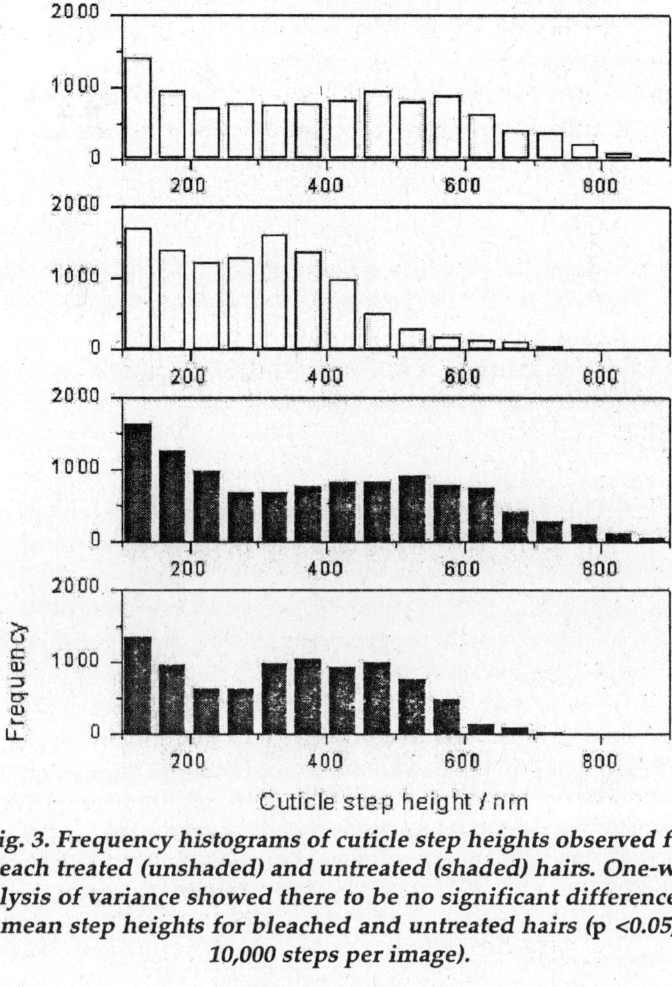

Fig. 3. Frequency histograms of cuticle step heights observed for bleach treated (unshaded) and untreated (shaded) hairs. One-way analysis of variance showed there to be no significant differences in the mean step heights for bleached and untreated hairs (p <0.05, N - *10,000 steps per image).*

Notes

1. Self-adhesive carbon discs (Agar Scientific, UK) tend to be more adhesive than carbon tape, and are especially suited for wet-cell work (although the experiments described here were performed in air).

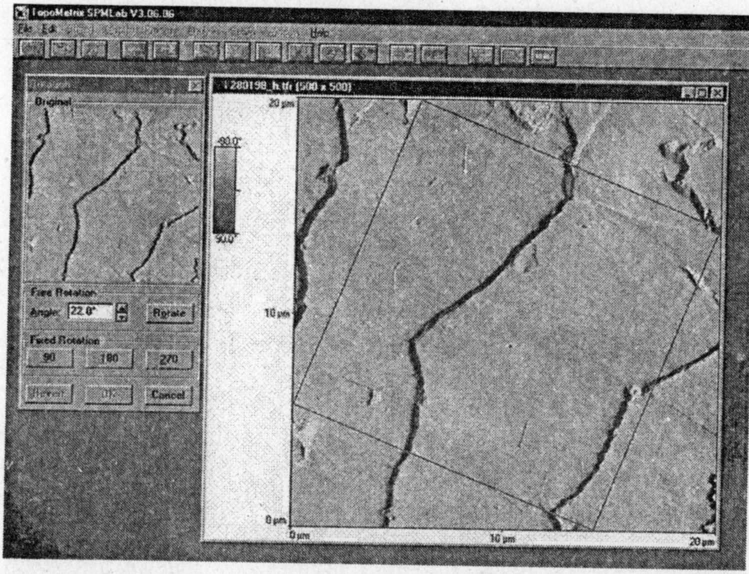

Fig. 4. *The cuticle step height program requires the scales to overlap from left (root end) to right (tip end). This can be achieved either by rotating the sample (preferred method), or, as shown here, rotating the image (dataset) through 22°.*

2. The best way of viewing the surface architecture of cuticle scales is with the orientation such that scales overlap from top–left (root end, highest point) to bottom–right (tip end, lowest). With the TopoMetrix Discoverer TMX2000 instrument, this can be obtained by mounting the hair perpendicular to the long axis of the cantilever and performing the scan at a scan rotation of either 90 or 270°, depending on juxtapositions of the root and tip ends. However, the required orientation for the cuticle step height program is for the cuticle scales to overlap from left to right. This can either be achieved by rotating the scan direction before acquisition, or rotating the image, typically by about 20°, during the image-analysis routine (Fig. 4).

The latter, less favorable method reduces the scan range of the image and so reduces the number of cuticle steps per image.

3. The program also assumes that the image is free from artefacts caused by the scanner going off-range as a result of the gross curvature of the hair sample (*see* Subheading 3.2., step 4). It is also necessary for the hair sample to be free from endocuticular debris, which is sometimes observed at the foot of cuticle steps, otherwise an erroneously low step height may be recorded.
4. The general program methodology is as follows: The program opens the input text file and examines the first line consisting of 500 data points of height values recorded in nanometres. The line of data is then derivatized using a first-order algorithm to locate the positions of the cuticle steps on the x axis. Markers are set that tag the start and end of each cuticle step on the line profile. This is achieved by defining a threshold gradient, currently set to –10 nm/pixel. The minus sign indicates that the cuticular sheets step down from left to right across the image; the units are simply a result of z data being recorded in nanometers and lateral information being recorded in pixels. The points where the derivative plot falls below the threshold and subsequently rise above it mark the start and end of the cuticle step, respectively. For this, the program considers all the points below the threshold gradient and measures the gap between the current pixel position and its previous value. If the gap is greater than a given clearance, currently set to 10 pixels (0.4 µm for a resolution of 500 pixels), then a new cuticle has been identified. The routine continues until the end of the line. Cuticle step heights are then calculated by measuring the vertical distance between the start and end markers for each cuticle step. Steps heights less than 100 nm or greater than 900 nm are neglected. The process is repeated for the remaining 499 lines of the input text file. The program can easily be adapted to open further text files corresponding to more hair images to construct a more representative sample set.

9

Imaging Living Chondrocyte Surface Structures with AFM Contact Mode

INTRODUCTION

In its most established mode of operation, named constant force contact mode, atomic force microscopy (AFM) has been applied to image the 2D and 3D architecture of surfaces. Any deflection of the tip as a result of surface topography is recorded. The microscope reconstructs an image of the surface from the x, y, and z scan data to develop a 3D illustration of any surface at the micro- and nanometer level.

The production of high-resolution images of a wide variety of biological samples at near-native conditions and the possibility to measure very low local forces is proving to be a powerful tool for cell analysis. In contrast with electron microscopy observations in particular, AFM improves biological studies involving imaging by also monitoring dynamic processes.

However, the investigation of soft biomaterials with this special method is still challenging. This chapter reviews practical details of imaging two cell lines: human chondrocytes and human osteosarcoma.

However, characteristics described are not unique to this type of cell. Principally, all types of adherently growing cells can be investigated with the techniques described here. Further sections explore how AFM can be used as a helpful tool in observations of the cell surface and the physical interactions that occur there, like adhesion or friction, and their influence on the active cell.

CELL LINES

Characteristics of Chondrocytes

Investigations were performed on human chondrocytes isolated from human osteoarthritic knee joint cartilage. The cartilage was isolated from cartilage bone fragments resected during the insertion of knee prostheses. All patients presented gonarthritis. No other relevant disease—particularly rheumatoid arthritis—was present. Immediately after the resection, the cartilage bone-fragments were potted in sterile L15 medium (Seromed, Berlin, Germany). Thereafter, the cartilage was handled as described elsewhere.

Cartilage is comprised of a large amount of functional extracellular matrix that is made and maintained by a small number of chondrocytes, the sole resident cell type. Chondroblasts and chondrocytes secrete cartilage matrix, and chondrocytes are also embedded therein. The bones of a developing or restoring limb form through the process of endochondral bone formation. In the beginning, mesenchymal cells condense and cells in the core differentiate into chondrocytes, and the cells at the periphery differentiate into the perichon-drium. Articular cartilage has several features that impact on the fate of bioactive bodies. Chondrocytes are anchored in the extracellular matrix and are surrounded by a pericellular matrix. Of particular interest regarding dense connective tissues, recent experiments have shown that mechanotransduction is critically important in vivo in the cell-mediated feedback among physical stimuli, the molecular structure of matrix molecules (e.g., collagen), and the resulting macroscopic biomechanical properties of the tissue.

Characteristics of Human Osteosarcoma

Human osteosarcoma (HOS), a human osteogenic sarcoma cell line, was purchased from American Type Culture Collection (Rockville, MD). The cells were cultured in a medium volume equivalent to 1:1 mixture of Dulbecco's modified Eagle's medium (DMEM) and Ham's F-12 medium containing penicillin (100 U/mL) and streptomycin (100 μg/mL) and 10 vol% fetal bovine serum. The HOS cells exhibit a flat morphology, low saturation density, low plating efficiency in soft agar, and are sensitive to chemical and viral transformation.

The nontumorigenic, as well as the immortal tumorigenic, osteoblast-like human osteosarcoma cells are used in many laboratories along with their large number of derivatives. Because they are one type of potential hormone-related cancer, the number of studies is incredibly high. For these cells to reach their functional differentiated state the action of specific factors is required.

Mechanical stress is an important regulator of bone metabolism. Fluid shear stress caused by mechanical load in bone tissue has been shown to be important to both the bone structure and function through its effects on osteocytes and osteoblasts. Many hypotheses about the mechanotransduction system in bone cells have been proposed. Recent findings suggest that the physiological level of fluid shear stress induces the production of crucial proteins in human osteosarcoma cells via the cation channel function and, as a result, may therefore promote bone formation.

AFM CONTACT MODE IN BIOLOGY

In an AFM the tip is mounted on the end of a flexible cantilever. As the sample is scanned beneath the tip, small forces of interaction with the sample cause the cantilever to deflect, revealing the sample's topography. The most common approach—called an optical lever—is to reflect a laser beam off the backside (upper side) of the cantilever into a four-segment photodetector (quadrant). The difference in output between the detectors is then proportional to the deflection amplitude. Important to note is that the limiting factor in motion detectors is not the sensitivity of the photodetector itself (deflections as small as 0.01 nm can be detected), but the intrinsic vibration of the cantilever attributable to Brownian motion.

The cantilever is integrated with a sharp tip on the end and characterized by its material (usually silicon nitride for contact mode investigations), its spring constant, and its geometric properties (usually parabolic or pyramidal tip shape with a curvature radius of 20–40 nm). The spring constant, k_n (determined by thermal vibration in air) varies from 0.06–5 N/m. Low spring constants are sensitive to uncontrolled vibration of the tip released by tip–sample interactions.

Contact Mode Description

In the contact mode, the tip touches the surfaces at all times with constant force, sliding over the surface as the sample is scanned line by line. Thereby topographic information is received by monitoring the change in cantilever deflection. Force-distance curves are obtained by plotting the vertical displacement of the cantilever, as a function of the separation between the tip and the sample. The force curve is an approach-retract cycle, in which the sample first approaches the tip (*see* Fig. 1) and is subsequently retracted from the tip. The cantilever deflection "z is then converted into force (F_n) according to the relationship:

$$F_n = k_N \text{ "}z$$

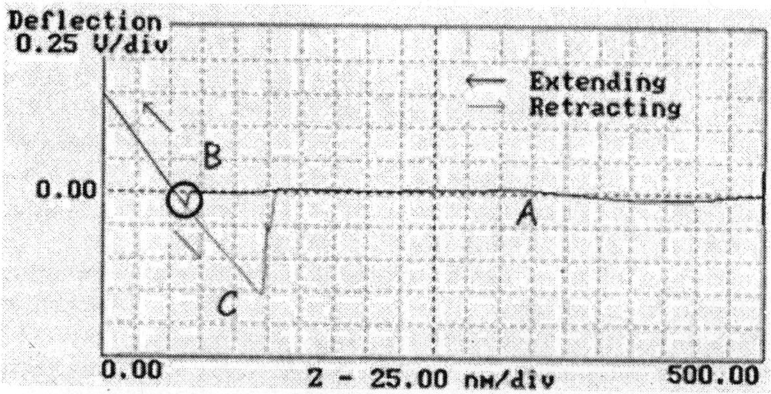

Fig. 1.* Favorable force–distance curve of an adherent cell. (A) No interaction force detectable at large tip–sample distances. The distance of the scanner movement is represented by the horizontal axis, and the cantilever deflection is represented by the vertical axis. In the case shown, there are minimal long-range forces, so this "noncontact" part of the force curve shows no deflection. (B) As the probe tip is brought very close to the surface, it may jump into contact (see circled area), if it feels sufficient attractive force from the sample. Sometimes repulsion force induces elongation in other directions. As the tip moves further in the positive z direction, a positive linear cantilever deflection is observed as the tip and sample move together. If the cantilever is sufficiently stiff, the probe tip is able to indent into the surface at this point. If this takes place, the slope of the contact part of the force curve can provide information about the elasticity of the sample surface. After loading the cantilever to a desired force value, the process is reversed. As the sample moves in the opposite (negative) z direction, a similar cantilever deflection line is traced as the tip and sample remain in contact. (C) As the tip moves further in the negative z direction, the restoring force exerted by the bending of the cantilever overcomes the adhesive force of the tip–sample contact. At this point, the adhesion is broken and the cantilever comes free from the surface. This can be used to measure the rupture force required to break the bond or adhesion.

Since normal spring constants for cantilevers are 0.01–100 N/m and instrumental sensitivities for normal deflections are up to approx 0.01 nm, the corresponding limits in force detection are 10^{-13}–10^{-8} *N (12)*.

Because of their softness, the biological membranes of viable cells become significantly indented upon contact by the AFM scanning tip, even at low forces. Always exercise caution when interpreting the topographic features, however, because of the convolution of the tip shape.

To this point, we have focused on imaging mechanisms that rely on deflections of the tip with respect to the surface normal. The force generated when the tip is moved laterally over the sample surface can also be used as an imaging mechanism (phase or friction mode). The energy differences in trace-retrace plots are indicative of the energy dissipated in the scan. Attractive and repulsive forces lead to information about the hydrophobicity and hydrophilicity of the specimen. Changes in the "friction" images indicate quite well the tip– surface interactions. This is also true of a tip with a truncated apex ratio—the lateral force is rather insensitive to minor cantilever stiffness, different from the topography scan. In the case of round massive cells, high lateral forces, however, still hamper stable imaging.

AFM INSTRUMENTATIONS

AFM investigations were done at room temperature in air (samples covered with a droplet of water) or in buffer solution. We used the commercially available Digital Instruments Nanoscope III in constant force contact mode. Generally, the 512 × 512 pixel images were captured with a square scan-size between 0.6 and 100 µm at a scan rate of 0.2–5 scan lines/second (s) (0.2–5 Hz). Sharp Si_3N_4-cantilevers, each with a pyramidal tip, were used. Their spring constant was 0.1–5 N/m. Best results were obtained by using cantilevers with a spring constant about 0.5–1 N/m. To avoid cell damage, the feedback set point was adjusted frequently to 0.1–10 nN in order to optimize the contact force.

AFM IMAGING CONDITIONS

The data were acquired simultaneously with the height, the deflection, and the friction signals (*see* Fig. 2 to distinguish between the modes). The height mode monitors the topography. The deflection mode, as the first derivation of the height mode, offers supplementary details of the cell structure. The friction signal was used to investigate the lateral force interaction between the tip and the sample.

AFM Si_3N_4 tips should in principle be oxidized and hydrophilic; however, in practice they will be hydrophobic owing to hydrocarbon contamination. Fluid imaging with AFM requires a special tip holder ("contact mode fluid cell" from VEECO Metrology Group [Mannheim, Germany] was used). For the microscopical studies, the chondrocytes and HOS cells were seeded onto round glass cover slips (4-mm diameter). These cover slips were attached to the bottom of the fluid cell with vacuum grease to standard magnetic AFM mounting plates, before being covered with some droplets of media.

When the tip dives into the liquid medium, the laser reflections have to be carefully inspected to exclude "false" reflections, which occur when the tip comes in contact with the liquid surface. Usually our measurements were done at room temperature in aqueous phosphate buffer solution. Good imaging and detection of cell activity could be obtained in the constant force contact mode. The cantilever was carefully approached to the surface), in order to collect the first images in the "low-contact" mode and to avoid strong physical contact between the tip and the sample surface.

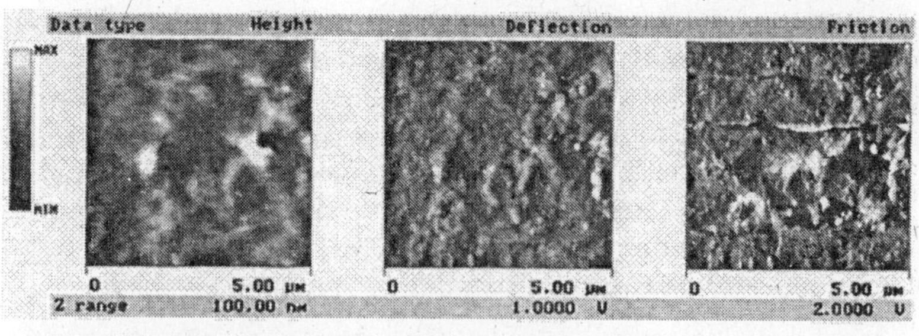

Fig. 2. Simultaneous AFM images of HOS cells in buffer observed with different modes: height (left), deflection (center), and friction (right). The topography is monitored by the height mode. The deflection mode, as the first derivation of the height mode, offers more details of the cell structure. The friction signal was used to investigate the lateral force interaction between the tip and the sample. Generally, large contrast in friction image often indicates active parts.

Later on, scans were done in contact mode with increased forces. Under favorable conditions, cells could be observed for up to 8 hours (h) depending on the cell viability. Frequently, undefinable cantilever vibrations induced by diffusion processes or cell motion are challenging problems. As a practical note, best observation conditions occur at night, when the neighborhood vibrations are minimized.

AFM CONTACT MODE IMAGING OF LIVING CELLS

Cantilevers with a spring constant have a reduced sensitivity to vibrations and are used successfully to surmount undefinable cantilever deflection. It is of great importance to adjust and minimize the force carefully and to avoid cell damage. In contact mode, true molecular resolutions could be achieved. The investigation of adherently growing cells with very low pressure on the tip resulted in diminished cell motion and improved the study. Well-resolved topographic information could be obtained. Zooming-in allows the recording of pictures with increasing detail. Especially in fluid medium, the investigation of active cells offers numerous facts. As an example of dynamic interactions, a series of images collected from chondrocytes and HOS cells in buffer is presented in Figs. 3–6).

High-resolution images of inner pore processes from the chondrocytes could be visualized (Fig. 3 and 4). During the pore diameter reduction, the surface potential in the immediate vicinity changes noticeably. The dynamic interaction is followed by secretion. Large differences (high contrast) in the friction images of several chondrocyte measurements (Fig. 4) point out an active part of the cell surface. The data was recorded during an interval of more than 2 h. The friction images remained a rather constant dynamic during this time. This is an indication of the viability of the material. This time interval of several hours seems long enough to study cell stimulations with mediators (e.g. cytokines, mitogens, enzyme substrates) and thus offers great promise for future experiments.

However, the round massive chondrocytes should not be as suitable for AFM observations as the flat HOS cells. In (Fig. 5). the secretion on their cell surface is monitored (see in particular some cell excrements marked out in frames A and B). In this case, we were able to obtain good-quality images quite easily and visualize the cell structure. A collection of force-distance curves could be collected in order to control whether the tip indented the soft cell surface or not. Indentation increases with the applied force and reaches a maximum value, after which tip-soiling damage occurs.

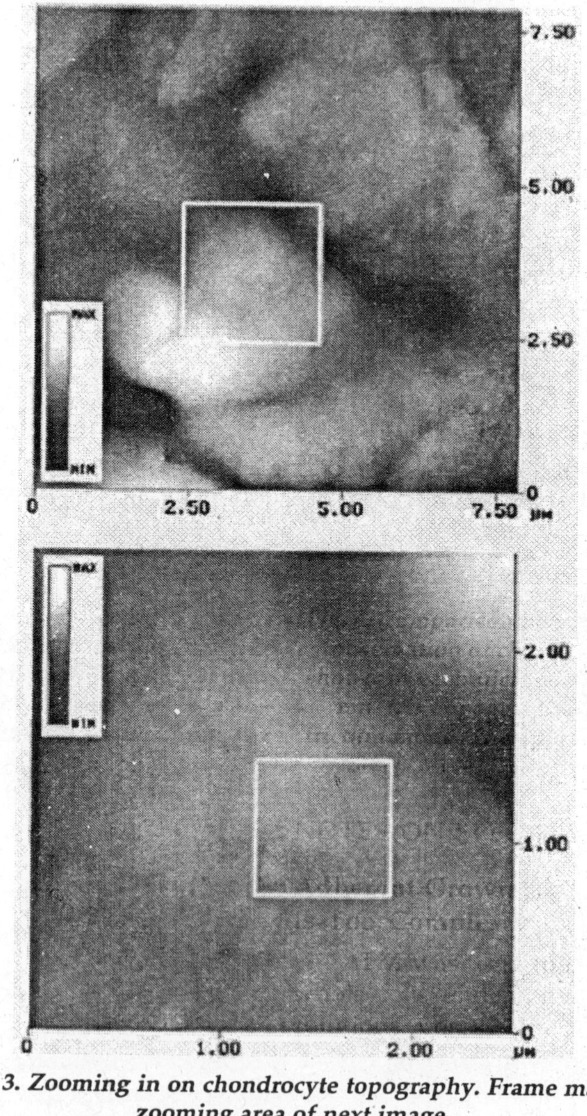

Fig. 3. Zooming in on chondrocyte topography. Frame marks zooming area of next image.

However, the surface penetration results in almost any case in more or less tip contamination. While the shape of the biofouled tip had broadened at the apex in comparison with that of the original tip, further investigation had to be done after replacing the tip. These effects and objections are described in more detail in Subheading 5.

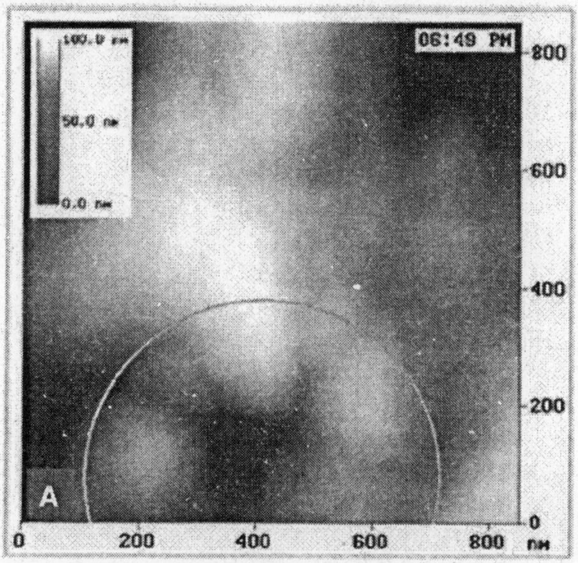

Fig. 4. ***Comparison between height and friction mode imaging. AFM observation of chondrocytes in buffer. The measurement time is shown in each picture. All images span an actual field of 850 × 850 nm. (A) Topography scan over 1 h simultaneous in height and friction mode. The approximate pore diameter is reduced from 382 ± 10 nm to 338 ± 10 nm.***

NOTES ON SPECIFIC DETAILS

Adherent Growing Cells Pose a Problem: Their Topography is Too Complexly Exhibited for Scanning

AFM was used to investigate different viable cells. Scanning whole cells under physiological conditions, in media or buffer solutions, poses some problems.

Since most cells are too large to observe them as a whole (Fig. 5), only portions of the cells can be investigated. Figs. 7 and 8A (pp. 117, 118) show rare examples of cancer cells that are small enough to scan whole. Numerous problems result in investigating the identical local position several times by AFM, after the material has left the Nanoscope instrument for other investigations. Molecular marker might help to solve this problem.

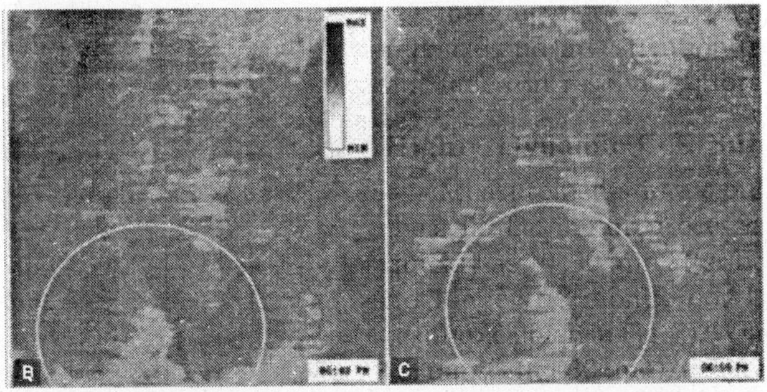

Fig. 4. (B,C) Large potential differences in the friction image of several measurements indicate an active part of the cell surface. During the pore diameter reduction, the surface potential in the immediate vicinity changes noticeably. A circle marks one active center on the cell surface. The timely changed contrast in friction mode between measurements indicates diminishing cell surface activity. This can be studied with much more detail by using different colors (15).

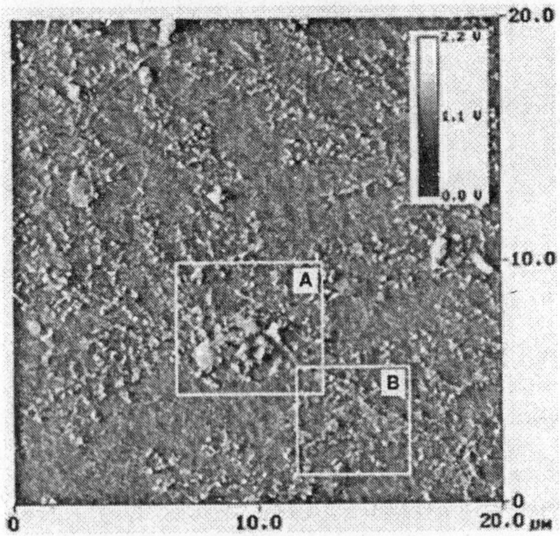

Fig. 5. Overview of the flat epithelial HOS cell in buffer observed in deflection mode. The secretion on the cell surface is monitored (some cell excrements in frame B are magnified in Fig. 6).

HIGH RESOLUTION IMAGING OF CELL SURFACES REQUIRES TIGHT

Attachment to Substrates

Imaging of loosely adhered cells enabled determination of the cell size and investigation of larger structures and pseudopodia but failed in resolving more detail.

Highly Dynamic Cell Surfaces Require Fast Scan Rates

Measurements in air only allow for limited examinations of the cells. Drying up processes strongly change the cell surface. As quickly as 10–30 min after beginning, dynamic interactions could no longer be monitored. Usually the cell structure collapses (example given in Fig. 7). When no dynamic changes were detected (usually after 30 min), the scan rate can be reduced from 5–0.2 Hz to increase image quality and resolution.

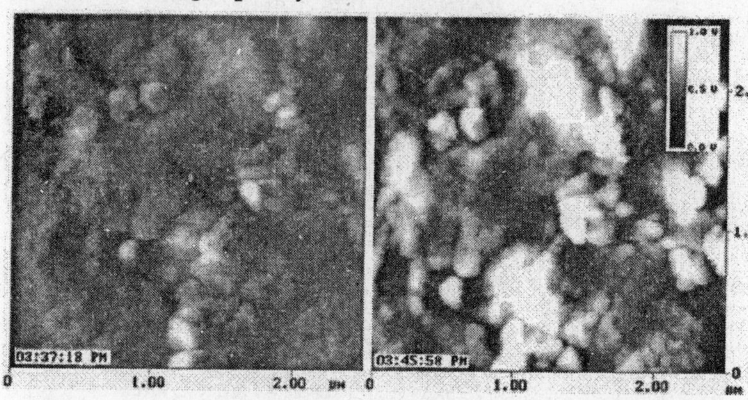

Fig. 6. Zooming in on frame B of Fig. 5 (same z scale for both figures). Increasing contrast in the deflection mode indicates growing roughness of the surface.

Fast scan rates in buffer induce several troubles (such as buffer turbulences) and undefinable cell vibration. Therefore, the scan rate is critical and dynamic interactions can only be monitored with restriction.

The Buffer Sometimes Crystallizes During Liquid Evaporation

Under the required conditions, the buffer frequently crystallizes during liquid evaporation. These crystals can be identified easily by their symmetric structures (phosphate buffer crystals are marked out in Fig. 8A).

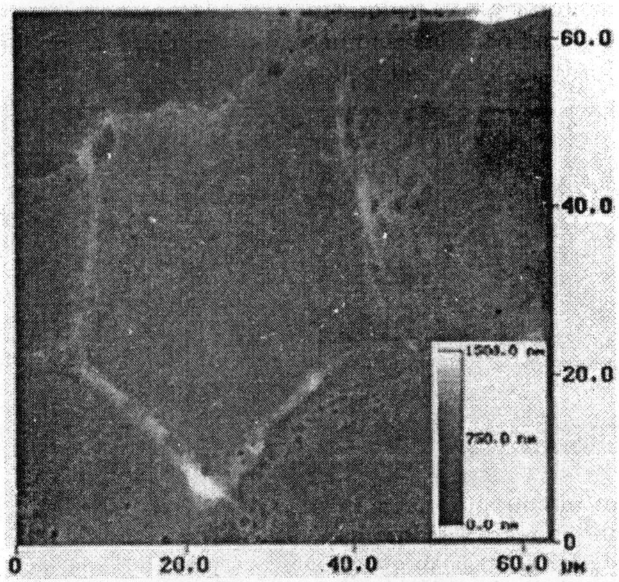

Fig. 7. Collapsed cell structure obtained in height mode AFM. The shape of the nucleus is visible. The cell margins (five white edges) are lifted on account of drying processes.

Protein Serum Covers the Cell Surface Like a Dense Carpet

The protein cover from a cell-culture procedure resembles to a high degree the protein surface of the washed cells, but it demonstrates a more homogeneous friction signal showing small changes with time. However, to distin- guish between the serum cover and the actual cell surface is at times very difficult (*see* Fig. 8B). Rinsing the cover slips with PBS several times before AFM observation is always necessary to remove the coating.

Soft Collagen Material Differs Clearly from Viable Cell Material

As shown in the force calibration plots of pure collagen material (Fig. 9), the tip–surface interaction can be discriminated easily from active cell parts by the shape of the curve. Approaching the surface over a long range (around several micrometer) leads to a low-force indentation of the tip into the moist and soft collagen matrix (Fig. 9A). No change in the shape of the force curve can be observed by retracting the tip. Therefore, it can be concluded that the tip has no attraction or repulsion interaction with the collagen sample. This can be used as reference to identify local collagen collectives (*see* Figs. 5 and 6).

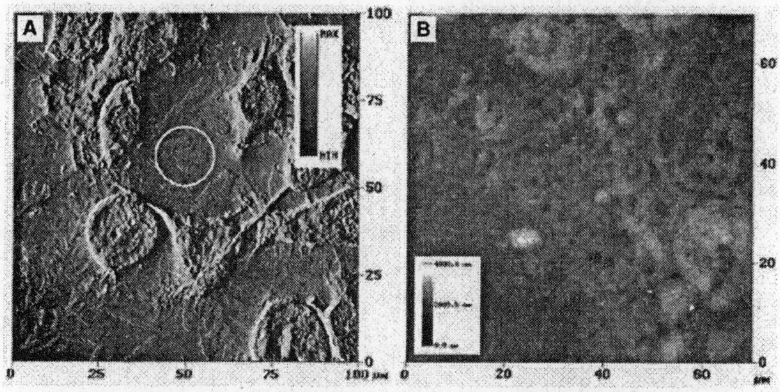

Fig. 8. Collection of frequent artifacts. (A) crystallization of phosphate buffer during liquid evaporation **(circle)*; deflection mode data. (B) Serum protein layer from cell culture medium covers the cells*** **(see** ***Subheading 7.5.); height mode data.***

When collagen dries up, it becomes a stiffer material with less depth and consequently it would receive a shallower indentation. An example is shown in Fig. 9B and C. The sample structure collapses after water evaporation (*see x*-axis). As the sample is approached to the tip, a strong positive linear cantilever deflection is observed after contact when the tip and sample move toghether. When the motion of the sample is inverted (retracting portion of the curve), a similar cantilever deflection line is traced, but this time shifted several nanometers due to tip and sample remaining in contact.

The Shape of the AFM Tip Is Always Critical in Sample Measurements

The quality of the tip (e.g., radius of curvature, morphology, hardness, and surface composition) influences strongly the quality of AFM investigations. Most commercially available tips have a curvature radius of 30–50 nm with a pyramidal geometry. (Several others alter the tip morphology to enhance imaging capabilities.) Additionally, the tip may bind proteins or membrane debris that diminish the resolution. Special coatings exploit this situation and make molecular interaction force measurements possible *(12)*.

ATTRACTION AND REPULSIVE INTERACTIONS OF THE TIP CAUSE

Misinterpretations, Especially With Viable Cells

Typically, when the tip approaches the surface, the deflection value increases to the set point, indicating surface contact. As soon as the tip is in contact with the cell surface, the active cell induces electronic signals by disturbing the scan. Sometimes, it seems that the cells are tickled by the tip (or the applied potential) and shake themselves. This movement overestimates the determined cell-size (Fig. 10; ref. *20*). Force-distance curves display uncontrolled vibrations of the tip. Undefinable surface tension forces could influence tip retraction (local attraction and repulsive forces could be detected).

By repeating the approaching procedure to physical contact, nearby the set point ("U about –0.1 V) the tip–sample distance jumps to higher values and evades a successful approach. Strong tip vibrations are observed. An explanation could be that the cell dodges or repels the tip (data not shown). Even after a while, viable cells seem to remember the procedure, because they induce strong tip vibrations and make repeated scans impossible. Moving to another part of the sample (tested up to a few millimeters) cannot prevent the repulsion. If such a phenomenon occurs, another successful approach and further investigation are impracticable.

To avoid such troubles the force and the cantilever sensitivity have to be optimized. A higher spring constant prevents the uncontrolled vibration of the tip, but results more often in cell damage. When the interaction between the tip and the surface is very strong, it seems very much like a scan done in honey or highly viscous material. Tip contamination often occurs. This is explained in detail in Fig. 11.

Cells Can Contaminate or Stick to the Scanning Tip

Occasionally during the scanning procedure, the tip is covered with an indefinite cluster. Tip-biofouling caused by cellular damage and pick-up of loosely adhered particles generate artifacts that will compromise the experiment. Since the tip geometry is critical for force measurements and accurate topography determination, this has to be avoided in all cases.

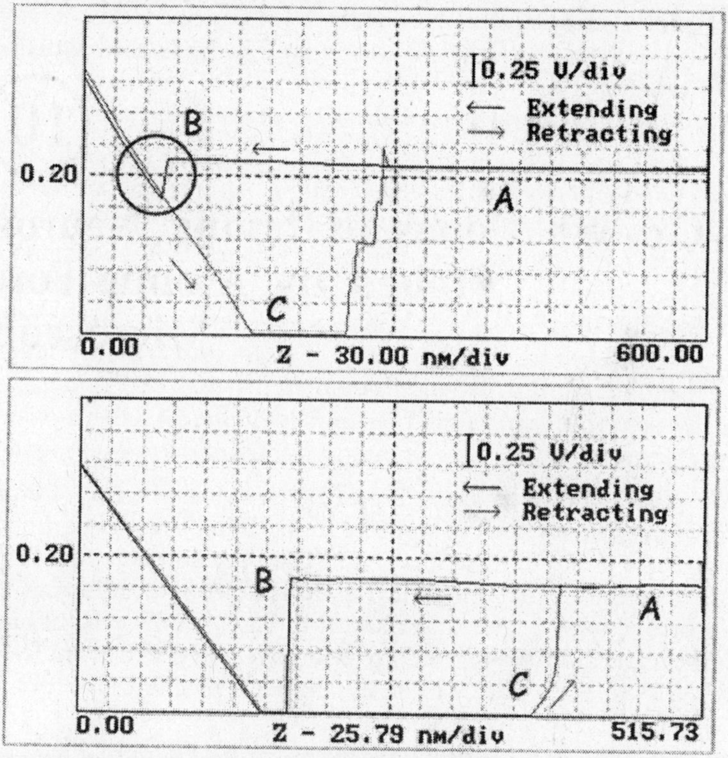

Fig. 11. ***Force–distance curve of a sample, which induce strong tip–sample interactions. Black line, approach curve; gray line, retraction curve. (A) At large tip–sample separations, there is no detectable interaction force. As the distance decreases, long-and short-range forces can be determined. (B) At some separation, the gradient of interaction energy exceeds the restoring force of the cantilever and the tip jumps to contact with the surface. If the interaction is too strong (bottom), no successful scan can be performed. (C) As the tip moves further in the negative z-direction (retracting), the restoring force exerted by the bending of the cantilever overcomes the adhesive force of the tip–sample contact and the tip breaks away from the sample. An idealized curve shows strong linear lines as shown in Fig. 1. The deflection curves presented here indicate tip contamination.***

10

Growth Cones of Living Neurons Probed by Atomic Force Microscopy

INTRODUCTION

A large body of recent literature describes the use of atomic force microscopy for the study of living cells. These experimental findings clearly indicate that AFM is a very valuable tool for the 3D imaging of flat biological samples strongly adhering to a substrate, with a lateral resolution in between the resolutions of optical and electron microscopy. Moreover, a very relevant feature of AFM is its capability of analyzing local mechanical properties of living cells.

The expression "flat biological samples" includes layers of cells, such as epithelia, and single cells, such as fibroblasts and glia cells. AFM technique, in its present state, seems to be less appropriate for globular structures, such as neuron bodies, and for string-like structures, such as neuron arborizations. However, neuron growth cones are subcellular structures that seem to be very appropriate for AFM analysis: they are flat, highly specialized regions, which make very strong adhesion to the substrate. Moreover, the mechanical properties of these structures (i.e., the cytoskeleton local organization) are of great relevance for understanding the development of neural architectures. The potential, therefore, of micromechanical information from AFM is of particular value.

On the basis of these premises, this chapter will be devoted to a detailed report of experimental findings concerning the use of AFM to probe growth cones of chick embryo spinal cord neurons under vital conditions

MATERIALS

1. Chick embryos (7–8 days (d) old).
2. Chick embryo extract (Gibco 16460-016).
3. Hanks' balanced salt solution (HBSS; Gibco 24020-091).
4. Bovine serum albumin (BSA; Sigma A-7030).
5. Trypsin solution 0.25% (Gibco 25050-014).
6. Trypsin inhibitor (Sigma T-6522).
7. DNAse (Deoxyribonuclease, type I; Sigma D-5025).
8. DMEM-F12 (Gibco 31331-028).
9. Fetal bovine serum (FBS; heat inactivated; Gibco 10108-157).
10. Horse serum (heat inactivated; Gibco 26050-070).
11. Poly-D-Lysine (Sigma P-7280) or Poly-L-Lysine (Sigma P-9155).
12. Stock supplement solution N-2 (Gibco 17502-048).
13. 5-Fluoro-2'-deoxyuridine antimitotic agent (Sigma F-0503).
14. Phosphate buffered saline solution (PBS; Gibco 14287-080).
15. Glutaraldehyde solution (Sigma G-6257).
16. Atomic force microscope: Park Scientific Instrument Autoprobe CP (Thermo-microscopes, Sunnyvale, CA).
17. Silicon nitride pyramidal tips on cantilevers with 0.01 N/m nominal spring constant (Thermomicroscopes, Sunnyvale, CA).
18. Dissection microscope (WILD-LEITZ).
19. Microdissection forceps (Fine Science Tools [FST] Dumon #5 biologie).
20. Forceps, large, small (FST).
21. Scissors, fine (FST).
22. Disposable conical tubes (Falcon 2170, 2195, or equivalent).
23. Disposable cell culture dishes (100 mm Ø Falcon, 35 mm Ø Falcon).
24. Phase contrast microscope (Diavert-LEITZ).

25. Thermo-controlled waterbath (37.5°C).
26. Centrifuge.
27. Glass slides.

METHODS

The methods described below outline: (1) the neuron cell culture and sample preparation, (2) the AFM setup for imaging, (3) the acquisition of force-vs-distance and indentation curves, (4) the results obtained and their interpretation, and (5) a comparison with other techniques.

NEURON CELL CULTURE AND SAMPLE PREPARATION

Chick Embryo Spinal Cord Neuron Extraction

Spinal cord neurons were obtained through dissection of spinal cords from 8-d chick embryos and plated on treated cover slips. Dissected cords were minced in HBSS and enzymatically dissociated in 0.05% trypsin at 37°C for 25 min, then washed in CMF-HBSS containing 0.3% BSA, 0.005% DNAse (deoxyribonuclease, type I), and 0.025% trypsin inhibitor. After mechanical dissociation, the resulting single cells were suspended in MEM-F12 (1:1) supplemented with 5% FBS, 5% inactivated horse serum, and 5% chick embryo extract for plating on culture substrata.

Neuron Cell Culture and Sample Preparation for AFM Investigations

To prepare the culture substrata, glass slides were first cut to 20- x 40-mm pieces and then cleaned and sterilized. They were then incubated overnight in a poly-D-lysin solution (5 mg in 50 mL of distilled water), rinsed three times in distilled water, and dried in a sterile hood. Plating was made on the glass slides, which were then placed into plastic Petri dishes.

The cultures were incubated at 37°C, 5% CO_2. Two days after plating, the medium was replaced with MEM-F12 96%, horse serum 3%, 1% stock supplement solution N2. To free cultures from non-neural cells, 72 hours (h) after plating an antimitotic agent (5-fluoro-2-deoxiuridine, 10^{-6} *M)* was added to the culture medium.

Cell Fixation

For the purpose of comparing results obtained on living cells, fixated cells were also prepared. In this case, after keeping cells for 4 or 5 d in culture, the medium was removed and cultures were briefly rinsed with PBS. Cells were fixed for 20-30 min using 0.8% glutaraldehyde in PBS. Finally, slides were rinsed twice with PBS and dried.

AFM SETUP FOR IMAGING

AFM Setup

A Park Scientific Instruments Autoprobe CP (Sunnyvale, CA) AFM was used, which was equipped with a scanner tube allowing 100 μm *(x, y)* maximum scan size and 6 μm *(z)* excursion. All experiments were performed using cantilevers with 0.01 N/m nominal spring constant and silicon nitride pyramidal tips.

Special care was taken to avoid contact between liquids and scanner, as this would cause permanent damage to the piezoelectric element and eventually to the high-voltage electronics. For this purpose, the top half of the microscope containing the scanner was enclosed in a polyethylene film sheet. This allows the scanner to move freely and does not interfere with the magnetic coupling of the sample holder. The cantilever chip was mounted on a chip holder that has a glass window behind the cantilever chip. To avoid air bubble formation, before mounting the chip holder into the microscope we wet the glass and cantilever chip with buffer solution from a syringe and allowed a droplet of water to be trapped (kept in place by surface tension) between the chip and the glass window.

The sample was then taken out of the Petri dish, with a film of buffer solution allowed to remain on the surface. To overcome the difficulties of gluing a wet glass slide to the sample holder metal disk and also to overcome the limitations of the *x–y* table that has only a 12- × 12-mm range, we used the following method. First, we fixed a whole glass slide with cyanoacrylate glue to the metal sample holder disk, which is then placed on the scanner as usual. Second, we placed Vaseline onto this glass slide and pressed the cell-covered glass slide firmly onto it. This allowed us to easily move the sample in search of a good area for imaging and also to quickly change it.

Tip to Sample Approach Procedure

The first step is to approach the tip to the sample as usual with the stepper motor until the drop hanging from the cantilever holder assembly meets the liquid covering the sample glass slide. A meniscus is then formed and from this moment the surface of the sample can be seen through the on-axis optical microscope. Tip-to-sample approach was always performed on a glass area next to the cell to be imaged, and before scanning the force setpoint was lowered to a small value (0.5 nN) to avoid cell damage.

AFM Settings for Imaging

Force-vs-distance curves before and after imaging were recorded routinely for cantilever deflection calibration purposes and for sample stiffness estimation. These curves have been transformed into force-vs-indentation plots, using as reference a force-vs-distance curve taken on glass during the same session. Images were taken with two simultaneous acquisition channels in the AFM: the z-piezo driving voltage and the error signal from the feedback loop. The first signal is proportional to the z-piezo displacement necessary to maintain the cantilever deflection (force) at the setpoint during scanning, whereas the second one records deviations of the cantilever deflections (hence from the set force) from the setpoint value.

To obtain imaging with higher spatial frequency resolution, we tuned the feedback loop parameters so that only the average cantilever deflection was kept near the setpoint value, allowing the system to generate a meaningful image from the error-signal channel, which has a wider frequency band. Typical scanning speeds were between 13 and 41 μm/s.

ACQUISITION OF FORCE-VS-DISTANCE AND INDENTATION CURVES

Force-vs-Distance Curves

Force-vs-distance curves were obtained by using the standard PSI software, which records the cantilever deflection, while driving the piezo in the z direction after a triangular wave. The software allowed us to set the wave frequency and to average the force-vs-distance curves taken consecutively at the same point. The curves corresponding to a given image were stored in a digital file

(1024 points for each force curve) for further processing. The force scale for these curves was calibrated by using, as a reference substrate, the glass the cells adhered to. Because the glass did not appreciably indent under the loads applied, from the slope of the linear portion (after tip contact) of the force-vs-distance curve we derived the conversion factor from the error signal (in mV) to the cantilever deflection (in nm) and hence to the applied force (in nN), through the spring constant K of the cantilever (Force = K × cantilever deflection, nominal K = 0.01 N/m). This conversion factor depended on the intensity of the laser beam reflected from the backside of the cantilever and on the area of the spot on the photodiode. Therefore, for each series of curves taken in the same session, we left the laser alignment unchanged and began and finished the experiment performing a calibration curve on the glass.

Force-vs-Indentation Curves

When pushed against a soft sample, the tip of the AFM will indent the surface and the shape of the indentation curve (i.e., the relationship between the load applied and the tip penetration) will give information on the stiffness of the sample. The force-vs-indentation curves were calculated by using the approach portion of the force-vs-distance curves. The first step was to take a force-vs-distance curve on a naked glass portion of the sample as reference. From this curve, the coefficient of linear relationship between the z-piezo displacement and cantilever deflection was derived. From each of the force-vs-distance curves taken on the cells the calibration line was subtracted, thus obtaining the force-vs-indentation curve (*see* Note 10).

RESULTS AND INTERPRETATION

Imaging

Figure 1 is a collage of various images (acquired in error mode) taken on the same growth cone of a spinal cord neuron adhering to a treated slide just taken out of the incubator. Figure 1A shows a topview rendering of the growth cone. Filamentous cytoskeletal structures are evident in the thick region (arrows). Figure 1B shows the top region of the cone (partially missing in Fig. 1A). Small dot-like structures are visible (arrows). A further zoom of the top right corner of the cone is shown in Fig. 1C. A meshwork of cytoplasmic structures appears (arrows). Finally, Fig. 1D shows the image of the growth cone after about 10 min of continuous scanning. The background globular structure on the left (arrow), present in both images, can be used to align the two images.

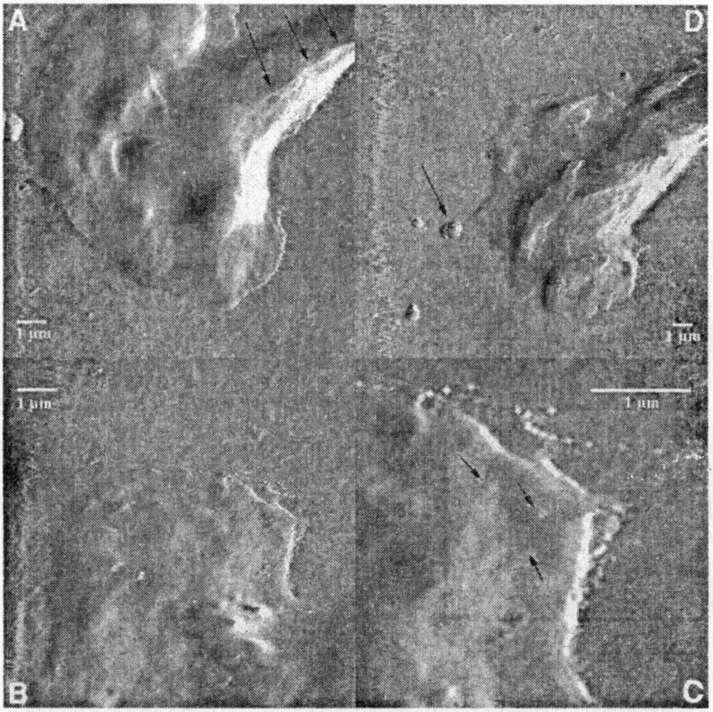

***Fig. 1.* Growth cone of a living spinal cord neuron adhering to a polylysine-coated glass slide. (A) Topview rendering of an error-mode image. Filamentous cytoskeletal structures are evident in the thick region (arrows). (B) Scan of the top region of the cone (partially missing in A). Small dot-like structures can be seen in the thick domain. (C) Zoom of the top right corner of the cone. A meshwork of cytoplasmic structures appears (arrows). (D) Image of the growth cone after about 10 min of continuous scanning. Most of the periphery of the growth cone has retracted.**

Most of the periphery of the growth cone has clearly retracted. An increase in the relief of the filamentous structures projecting towards the neurite can be noticed. Figure 2 shows a 3D rendering of the growth cone, as derived from the z-piezo (topographic) image (not shown), taken simultaneously with the image in Fig. 1A. It should be noted that the cone thickness shown in the figure is affected by the indentation of the tip on the neuron. Nevertheless, "true" thickness can be estimated and is described in Subheading 3.4.2. In the 3D image, a thick and a flat region can be tentatively identified, separated by a continuous relief.

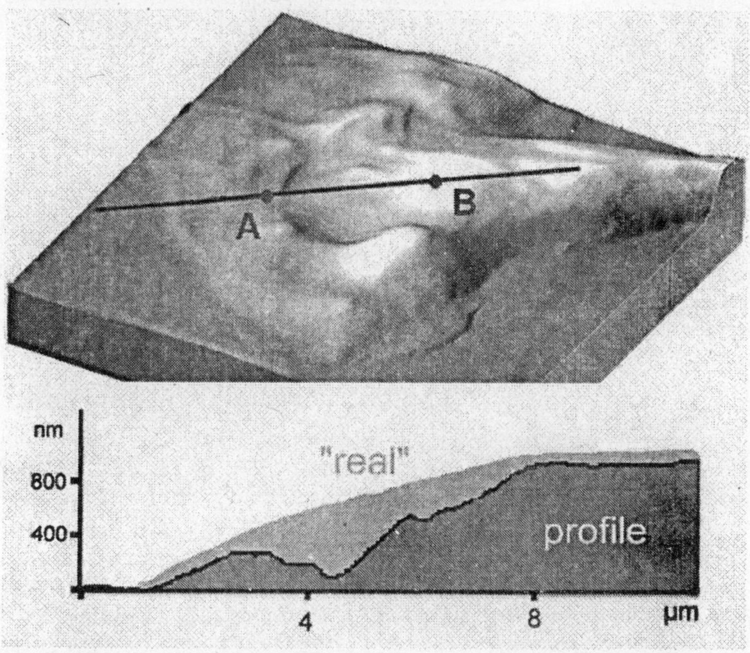

Fig. 2. 3D shaded rendering of the z-piezo signal image acquired simultaneously with the image in Fig. 1A, with a pictorial representation of the possible real-vs-mea-sured profile on the growth cone.

Figures 3A and B show the growth cone of another neuron analyzed immediately after leaving the incubator. A thick tubular zone is again evident towards the neurite. Careful inspection allows one to detect a surrounding low-contrast region with flat protrusions (arrows).

For comparison, Fig. 3C shows a similar growth cone after fixation. Similarly to Fig. 3A, Fig. 3D shows a growth cone from another living neuron, in which one can identify a thick tubular region surrounded by spiky structures (arrows).

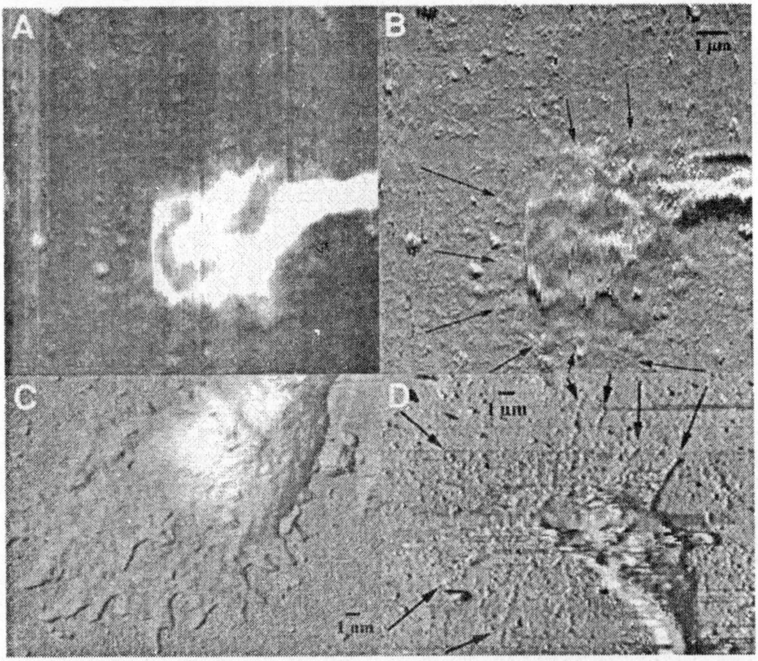

***Fig. 3.* Series of three images of different growth cones, in which the peripheral region has been detected by the AFM. (A,B) Growth cone of a living neuron analyzed immediately after leaving the incubator. A thick tubular zone is again evident. Careful inspection allows one to detect a surrounding low-contrast region with flat protrusions (arrowheads). Images obtained recording the z-piezo signal (A) and the error signal (B) simultaneously. (C) Image of a similar growth cone after fixation, shown for comparison. Z-piezo signal image. (D) Growth cone from another living neuron in which one can identify a thick tubular region surrounded by spiky structures (arrows). Image obtained recording the error signal.**

Figure 4A shows a small whole neuron with several arborizations. Towards the apical end most of them seem to be disrupted. Interestingly enough, a "trace" of the borders of the arborizations is evident (Fig. 4B and C). The trace is made of small (150 nm in diameter) dot-like structures, which could be identified as clusters of adhesion molecules.

Indentation, Topography, and Mechanical Properties

Figure 5 shows a series of representative force-vs-indentation curves acquired upon a growth cone of a living neuron.

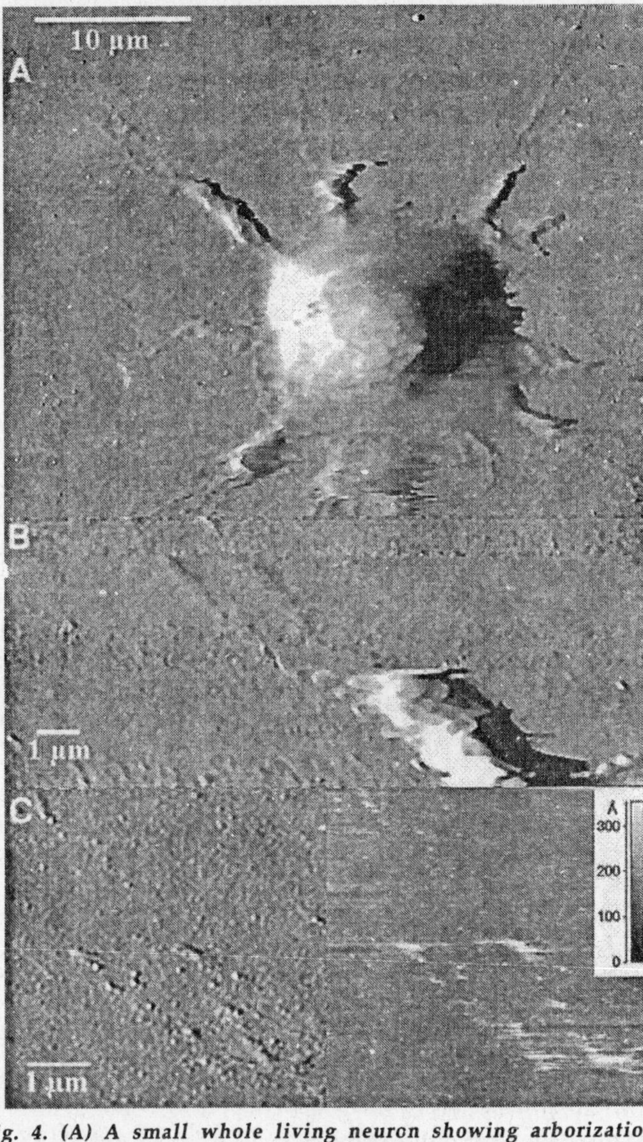

Fig. 4. (A) A small whole living neuron showing arborizations, imaged acquiring the error-channel signal. At the apical end most of the arborizations seem to be disrupted. (B) Higher magnification error-signal image of the apical end of an arborization. (C) Error signal and simultaneous z-piezo signal image of the same arborization apical end. The trace is made of small (approx 150 nm in diameter) dot-like structures that may be clusters of adhesion molecules.

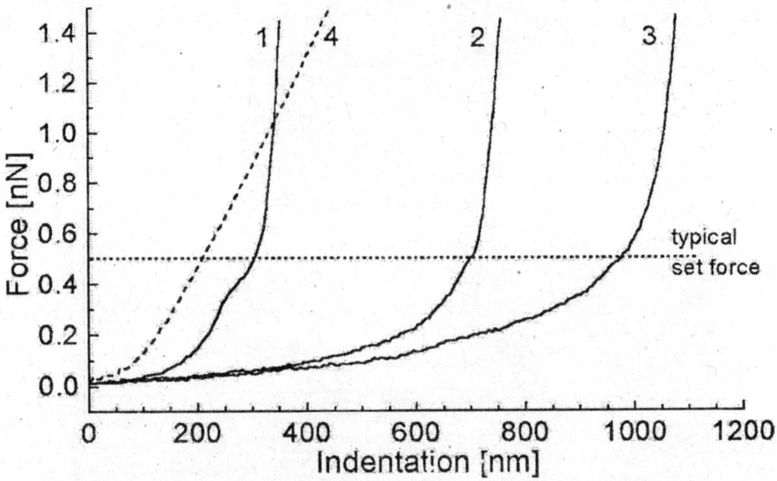

Fig. 5. Series of representative force-vs-indentation curves acquired upon a growth cone of a living neuron. Curves 1, 2, and 3 are taken moving away from the growth cone edge towards the neurite at steps of 3 μm. Each curve shows an increase in indentation with the applied force with a parabola-like behavior, until a quasivertical trend is reached. Curve 4 was taken next to the point were curve 3 was acquired but on a protrusion. A totally different trend can be observed: after an initial parabolic indentation a linear dependence on the force increase is established.

Curves 1, 2, and 3 in Fig. 5 were taken moving away from the growth cone edge towards the neurite at steps of 3 μm and represent the typical behavior of a "soft" portion of a living cell. The indentation at first shows a parabolic-like trend followed, at higher applied forces, by a quasivertical slope. This can be explained for the first part with the classical indentation theory of a solid punch into a half space, progressively deviating from such behavior as the glass substrate contribution becomes dominant. The quasivertical trend with increasing force indicates that the maximum compression of the cell material has been reached and the indentation limit value attained will give an indication of the cell thickness.

Let us now compare curves 3 and 4: curve 3 was taken onto an apparent depression of the surface, at a point like the one identified by A in Fig. 2, whereas curve 4 was recorded onto a stiff portion of the cell surface like the one identified by B. It is evident how in the case of curve 3 we are progressively indenting a thick portion of the cell until we reach the glass substrate, while in the second case, after a parabolic behavior for the first 100 nm of indentation, we reach a constant slope corresponding to an elastic spring constant of 0.0045 N/m. This means that, after indenting the most external "soft" cell surface, the tip interacts with submembrane structures that exhibit an elastic response *(4)*. At the force value of 0.5 nN that corresponds to the nominal setpoint used during imaging in the case of curve 3, we have an indentation of 950 nm, and for curve 4 we find 200 nm.

We need to keep these features and figures in mind in order to understand the contrast mechanism of both the z-piezo signal and error-signal images on such specimens. This means that the deep shallows next to high peaks found in the z-piezo image cannot attributed solely to morphological features. In fact, because on thick and "soft" locations, indentations can reach the micrometer range and on stiffer ones it falls to the 100 nm range, the shallows and peaks in the "topographical" image must be essentially the result of differences in stiffness of the submembrane growth cone structure encountered by the cantilever during scanning.

A pictorial representation of the possible "real"-vs-measured profile on the growth cone is shown in the second half of Fig. 2. Similar effects are found in the error images, where the gray scale levels represent the deviations from the feedback force (cantilever deflection) setpoint. Because the feedback loop has been tuned to give a high contrast in the error image, thus allowing temporary and relatively large deviations from the force setpoint, on the left hand side (scanning is from left to right) of an upwards slope or an increase in stiffness, the pixels will be darker, while on the right hand side they will be lighter. This can be clearly observed in Fig. 1A. Generally speaking, it is not possible to discriminate between a topographical change and a variation in stiffness, unless one has independent knowledge of the properties of the surface. By use of force-vs-indentation curves it is possible to discriminate the effects and estimate the thickness of the undeformed surface, at least in the point were the curve is taken. Extrapolation to similar areas can be made by an estimate of numerical values. A feature common to Figs. 1 and 3 is a thick tubular region extending towards the neurite. The thickness of this region as read on the z-piezo image is in the order of 1 µm, to which at least 200 nm must be added to take into account indentation.

IDENTIFICATION OF GROWTH CONE REGIONS

LAMELLIPODIA

The filaments in evidence in Fig. 1A and D can be easily identified as microtubules. By comparison with images generated by the other techniques and described in the literature, the thick region can be easily identified as the so-called C domain. Note that submicrometer size structures are visible in this region. In Figs. 1–3, this domain is surrounded by a flat area, with thickness in the order of a few hundred nanometers. This could be identified as a lamellipodia-rich P domain.

The flat protrusions shown in Figs. 3A and B can be identified as lamellipodia structures. Their thickness is in the order of 30–60 nm. Irregularities, distortion along the scanning direction, and "islands" of biological material underline the extent of the tip–sample interaction. For comparison, Fig. 3C shows a similar growth cone after fixation. Lamellipodia structures with a smooth profile are now evident. The thickness is now in the 100–200-nm range: the fixation process has affected the membrane stiffness so that negligible indentation occurs during scanning.

FILOPODIA

The spiny protrusions surrounding the C domain in Fig. 3D can be identified as filopodia structures. Interestingly enough, these protrusions appear to be made of globular subunits, often arranged in a discontinuous way.

These subunits have a diameter of 120–180 nm and a thickness ranging from 5 to 30 nm. They could be identified as clusters of proteins or patches of membrane adhering to the substrate, left after the tip–sample interaction.

The distribution of proteins in the filopodia of growth cones is a subject of active research. Filopo-dia are known to be filled with bundles of actin filaments, and the presence of spots of tyrosine-phosphorylated proteins have been recently demonstrated at the tips of growth cones by immunofluorescence techniques.

The formation of focal contacts by the tip of filopodia with the substrate is still an open question and further investigation of the described structures could contribute to answer it.

ARBORIZATIONS

Finally, traces of discontinuous biological material are evident in the terminal regions of the arborizations of a whole neuron (Fig. 4). A fixed similar neuron is shown for comparison (Fig. 6). Here, the arborizations are smooth and continuous. The morphology of the biological details is better preserved, but no information about adhesion to the substrate can be inferred. On the contrary, we can conclude that the interaction of tip with living material does somehow affect the morphology but, at the same time, gives hints at the nanometer scale about the organization of the biological structure and about the way contact is made with the substrate.

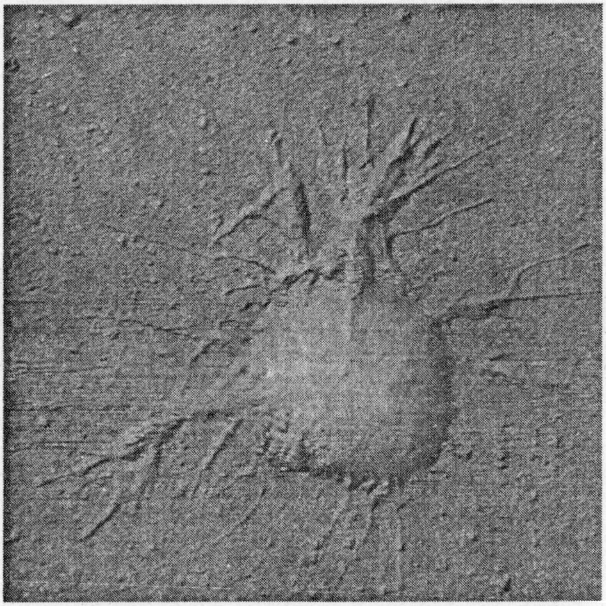

Fig. 6. A small whole fixated neuron showing arborization, to compare with the living one imaged in Fig. 4.

Comparison With Other Techniques

Detailed AFM images of living flat cells, such as glia cells, fibroblasts, and epithelial cells have already been analyzed in the literature. Low-resolution images of whole neurons have also been produced. The other available techniques for studying growth cones are as follows.

1. Whole-mount electron microscopy, which gives images with detailed information down to the nanometer but on dead materials and without thickness quantification.
2. Fluorescence microscopy, which allows one to identify cytoskeleton components by immunofluorescence staining. Time-lapse analysis of stained (lipid probe Dioc6) living growth cones have also been described.
3. Video-enhanced differential interference contrast (DIC) imaging, which is widely used for generating detailed images of unstained living growth cones. This technique has allowed the identification of two distinct domains: a central, relatively thick, organelle-rich region (C domain) and a peripheral, thin, region devoid of organelles (P domain).

AFM shares with the last technique the capability of imaging unstained samples. Moreover, as compared with all the mentioned methods, it is the only one to have the potentiality of giving quantitative information on thickness. However, to put the last statement in the correct perspective, it should be underlined that any AFM-originated image of a soft sample is the result of a mechanical interaction between tip and sample. This implies indentation, and force-vs-distance curves must be used to correct the data and obtain actual thickness. With indentation, a challenge to the adhesion of the living cone to the substrate is exerted. Finally, it is worth mentioning that, in principle, information about viscosity could also be obtained, similarly to the laser-tweezers technique, by carefully comparing the forward and retraction portions of the force-vs-distance curves.

Notes

1. Cut glass slides are used because the AFM employed for these investigations is a "scanned sample" model, able to accommodate only flat samples having maximum width of 25 mm and length of about 50 mm. When using "scanned tip" instruments, it is possible to apply the AFM directly onto the Petri dish. One advantage of using glass slides is their surface flatness with respect to Petri dishes.
2. Even if for some cell lines it is possible to grow cultures on bare substrates, for AFM investigation a good adhesion with the substrate is essential, allowing the cells to

withstand the lateral forces induced by the tip during scanning

3. An advantage of using small glass slides is the possibility of preparing several samples in the Petri dish at one time that can be kept in the incubator until just before use. This allows one to increase the throughput of one single primary cell line culture both in time and number of observable cells, as usually it is not possible to maintain temperature and CO2 control during AFM measurements. Recently, research groups have started developing systems that will allow one to control physiological environmental conditions during AFM imaging.
4. An important feature when using the AFM on living cells is the range available for the z direction in the scanner, as variations in height of several micrometers can be found during scanning. Also, x and y ranges should be several tens of micrometers.
5. Small spring constants avoid damage to the cell surface or even detachment of the cell from the substrate. We have also tried using 0.003 N/m spring constant cantilevers, but the adhesion forces between tip and sample did not result in good imaging.
6. Standard silicon nitride tips are quite sufficient for good imaging in contact mode on living cells as the visco-elasticity of the cell is the main limiting factor in resolution: in fact, a sharper tip does not improve resolution but can produce damage. Moreover, silicon nitride seems to behave better that silicon or polysilicon with respect to tip-sample adhesion.
7. A fresh supply of the buffer solution used for the culture, possibly held at 37°C, should be kept at hand, in order to be able to add it, when missing.
8. An on-axis microscope is essential for positioning the tip on the sample in the required position. Either a scanned sample AFM equipped with a high magnification, long focal length microscope or a scanned tip AFM mounted onto an inverted microscope can be used. The latter obviously opens the possibility of using a wider range of optical microscopy techniques in conjunction with AFM.

9. One of the serious limitations of AFM is the low scanning speeds that have to be used on soft surfaces. A 10 x 10-um image with 512 points per line is typically acquired in 8 min.
10. On the thinner portions of the cells, one observes at first the indentation process, and then a constant relationship between load and z-piezo travel is found. This means that all of the cell has been compressed and the glass surface has been "reached." One can use this last linear portion of the force-vs-distance curve to derive the coefficient and obtain the corresponding force-vs-indentation curve.

11

Evaluating Demineralization and Mechanical Properties of Human Dentin with AFM

INTRODUCTION

Atomic force microscopy (AFM) is a valuable technique for the study of demineralization and the effects of other solutions and environments on the structure of human dentin because high-resolution studies of changes in structure and dimensions are possible in nearly any environment over time. Because dentin forms the bulk of the tooth and is subject to a variety of alterations as a result of disease, age, and treatment, such studies are useful for understanding the basic structure–property relations of dentin and for evaluation of various conservative and restorative dental treatments. Initial studies of demineralization and dehydration of dentin were reported in 1993 in which relative changes in dimensions were used to evaluate these processes. It soon became apparent that a stable reference layer on the surface or embedded in the specimen would be most desirable for evaluation of time-dependent changes and to take advantage of the high-resolution capabilities of the AFM.

In 1995, the use of paint-on varnish or photoresist and evaporated gold layers were developed that allowed evaluation of demineralization, either continuously or sequentially, for a variety of demineralizing agents. An additional technique of interest is the use of AFM-based nanoindention for measurements of site-specific hardness and elastic modulus of dentin. Such measurements can frequently be performed as a portion of the same study in which demineralization experiments are conducted. The development of indentation-based mechanical properties measurements stems from the work of Doerner and Nix and can be performed with a variety of equipment types. We have been most interested in modified AFM methods because the AFM offers the possibility of both site-specific indentation and high-resolution imaging in solution or on dry tissue, whereas many other methods allow only the study of dry tissue.

Our initial work with AFM-based nanoindentation used a modified and stiffer cantilever and diamond tip that allowed hardness measurements to be made on either peritubular or intertubular dentin. However, modified transducer heads are now available that can measure load and displacement independently and provide site-specific load displacement curves and imaging in many environments. They offer substantial improvement and allow the determination of both hardness and indentation modulus (or reduced elastic modulus). Thus, we use this method in all current measurements of hardness and elastic modulus for dentin.

Continuous and Sequential Longitudinal Studies and Reference Layers

Continuous scanning is conducted by obtaining images continuously during demineralization in the wet cell of the AFM. Thus, each pixel of a line and each line of a scan represent different exposures and offer a continuous record of the demineralization process. In sequential scanning, a baseline image in water is obtained, and then the sample is removed, treated for an appropriate exposure time, washed, and reimaged in the AFM wet cell. This method has the advantage that start and stop times can be more easily controlled and each image represents a single exposure time so that changes can be measured on all the structures in an image.

Neither the varnish nor the evaporated gold procedures were ideal for establishing a stable reference area. For the varnish method, the thickness can be difficult to control, and if the thickness is too great, it makes a poor reference. In addition, the edge is usually rounded so that an abrupt step is not apparent and it is not always clear when the top surface is flat. Having a flat reference is useful when image processing, such as plane fitting, is used. Finally, the edge where varnish and dentin meet is the area that is used for measurement. This edge is often difficult to demineralize uniformly so that a typical image may have a rounded reference layer and a zone of uneven etching in the image field of interest, as shown in Fig. 1.

Thin evaporated gold layers (approx 10 nm) overcame many of these problems and allowed measurement between the gold layer and intertubular and peritubular dentin structures. However, there were two remaining problems with this method. Firstly, the dentin had to be subjected to a high vacuum treatment during gold evaporation. This may be undesirable in a naturally hydrated tissue and introduces microstrains. Second, demineralization for long periods in dilute solution or with more concentrated acids resulted in etching under the edge of the gold layer. When this occurs, the gold area is no longer a stable reference and any measurements are subject to greatly increased error.

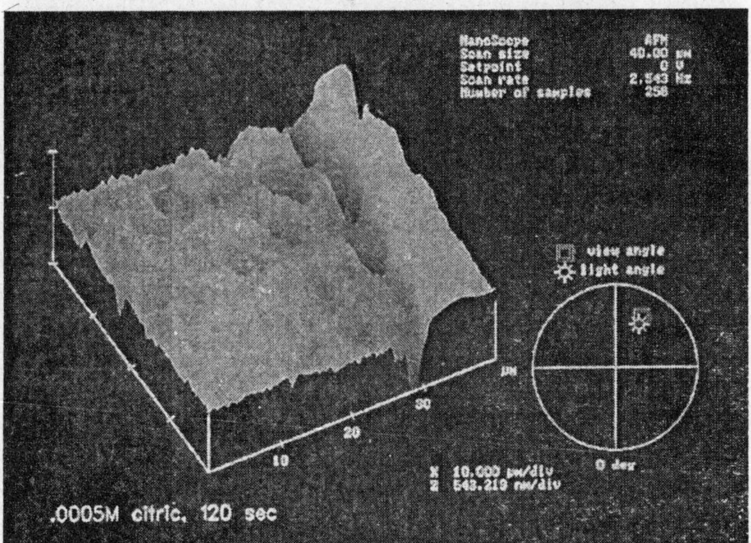

Fig. 1. Image of uneven etching that can be caused by a rounded edge of varnish, or the inability to apply the demineralizing agent uniformly at the edge of the reference layer. The etching time was 120 s with dilute citric acid. The reference area is in the upper right. A deep trough is adjacent to the layer and the uneven etching decreases with distance from the reference area, so that it is not possible to tell which areas of the dentin are representative of the etching exposure.

Three improved methods have been developed that can be used alone or in combination to overcome many of the preceding limitations. These include the use of an embedded cyanoacrylate reference layer that has been useful for demineralization studies in many dilute solutions as well as dehydration and rehydration studies and an embedded glass reference layer that has been developed for studies in which the cyanoacrylate method is unstable. The third method is to use a tape mask to protect portions of the sample from the solution, allowing the protected portion to serve as the reference area. This has been useful for studies of single exposures to concentrated solutions and the effect of drying and rehydration. We recently have used the combination of a mask and embedded layer to evaluate the combined effects of dem-ineralization and deproteinization of dentin.

Testing the Reference Layer Stability

When evaluating the effect of any new solution, the first step is to determine the stability of the reference layer. For example, cyanoacrylate is evaluated by bonding glass slides together with the cyanoacrylate, embedding in acrylic, followed by sequential metallographic polishing of the glass– cyanoacrylate–glass interface through a 0.05-μm alumina slurry. The sample is then imaged in the AFM, exposed to the solution of interest for appropriate periods, and then reimaged to determine any changes in height of the cyanoacrylate relative to the glass. For many solutions that we have studied, the cyanoacrylate has proven to be stable for exposure periods of at least 30 min.

We have found that for some solutions, such as ethanol and acetone, the cyanoacrylate is not stable, and this has led to the development of a glass reference layer method that requires additional steps for fabrication as described elsewhere. For dentin samples that have been etched extensively and then undergo dehydration, the interface between reference layer and dentin is damaged and often fails because of the drying stresses. To overcome this problem the extent of the etching must be limited so that the interface withstands these stresses, or a masking technique can be used, in which part of the sample is protected from the acid and the unetched portion serves as a reference area. This method has been used in several studies and only requires the identification of a tape that can be applied to the dentin and removed without leaving a residue. One such tape is Scotch Mounting Tape (3M, Minneapolis, MN). It also should be noted that the methods described here could be used for other calcified tissues. Fig. 2 shows an example of enamel etched to reveal the enamel prism structure with an embedded cyanoacrylate reference layer.

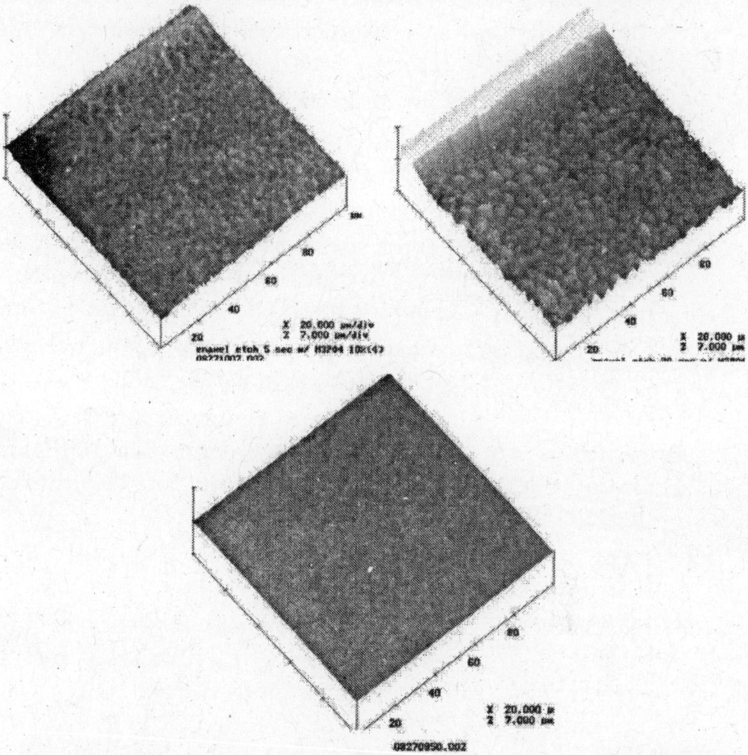

Fig. 2. Enamel etching sequence using the cyanoacrylate reference-layer method. Images for etching exposure to 10% phosphoric acid are shown at 0, 5, and 30 s. Also note that the etching pattern changed from one type to another during this treatment. The reference layer along the upper left of each image.

MATERIALS

1. Obtain teeth from human subjects following protocols approved by Institutional Review Boards and with informed consent.

2. Store whole teeth in filtered and purified water or Hanks' balanced salt solution (HBBS) at 4°C.
3. Teeth are potentially infectious and therefore should be disinfected or sterilized. Normally, we sterilize using low dose gamma-radiation.
4. Tooth sectioning is conducted using a low-speed water-cooled diamond saw (Isomet Low Speed Saw, Beuhler, Ltd., Lake Bluff, IL).
5. Store cut sections in HBBS at 4°C.
6. Sequential grinding is done using sand paper from 240, 320, 600, and 1200 grit (Buehler Ltd.).
7. Final polishing is performed using aqueous suspensions of alumina powder or diamond (Buehler Ltd.). A fine polish will leave the structure with no detectable smear layer. Ultrasonic clean for 10-15 s between polishing steps.
8. Embedded reference layer is made from ethyl a-cyanoacrylate (MDS Adhesive QX-4, MDS Products, Anaheim, CA).
9. Solutions are usually prepared from reagent grade chemicals except for evaluations of commercial demineralization agents.
10. AFM imaging performed with a Nanoscope III (Digital Instruments, Santa Barbara, CA).
11. Nanomechanical properties (hardness and reduced elastic modulus) are ascertained with a Triboscope indentor system (Hysitron Inc., Minneapolis, MN).

METHODS

Basic AFM Mode

AFM methods are well established. For studies of dentin recession and demineralization, we normally use the contact mode and standard S_3N_4 tips. Tapping mode also can be used and allows the tip force to be reduced. When coupled with high-aspect ratio Si tips with a very small tip diameter, it is most useful for high-resolution studies of particular features of the dentin. However, most steps of the demineralization studies do not require the high-resolution tips, which are expensive and fragile, and, therefore, the less-expensive standard tips and contact mode are used for most studies. Because dentin is a naturally moist hydrated biological composite, nearly all imaging is done in a wet cell filled with purified water at ambient temperature.

Dentin Disks With Embedded Reference Layer

1. An extracted human tooth is usually the sample of interest for studies of dentin. Because the tooth is a potential source of infection, after extraction we store the teeth in vials in distilled water or HBBS and sterilize with a low dose of gamma radiation. The tooth in its solution is then stored at 4°C until prepared.
2. The tooth is removed from its storage vial, cleaned of any soft tissue, and mounted on a wooden tongue depressor with hot glue. Portions of the tooth that are not of interest are usually cut off (e.g., pulp chamber and roots), and then the crown is sectioned longitudinally (sagittally) into two halves, as shown in Fig. 3A, using a water-cooled diamond cut-off saw (Buehler, Ltd Isomet Low Speed Saw, Lake Bluff, IL).
3. The cut surfaces (Fig. 3B) are then carefully polished through successive grits of sand paper (240-1200 grit) under flowing water, followed by aqueous suspensions of alumina in successive steps (1 and 0.3 um) and ending with 0.05-um alumina. Between steps, the sample is ultrasonically cleaned in distilled water for 10-15 s to remove remnants of the abrasives. Aqueous suspensions of diamond are another good alternative.
4. The two polished halves are then blotted dry and bonded back together (Fig. 3C) with ethyl á-cyanoacrylate (MDS Adhesive QX-4, MDS Products, Anaheim, CA) using medium finger pressure (Note 3). The viscosity of the cyanoacrylate can be adjusted by air drying to increase viscosity or thinned by mixing with acetone.
5. Sections for AFM study then can be obtained by sectioning the reassembled tooth parallel to the occlusal surface using the same water-cooled diamond cut-off wheel (Fig 3D). Because the dentin structure varies with intratooth location *(1)*, it is often of interest to prepare disks representative

of either superficial dentin (close to the enamel) or deep dentin (close to the pulp chamber). Usually one or more disks of 1- to 1.5-mm thickness are obtained from a single tooth by repeated sectioning parallel to the occlusal surface (Fig. 3D).

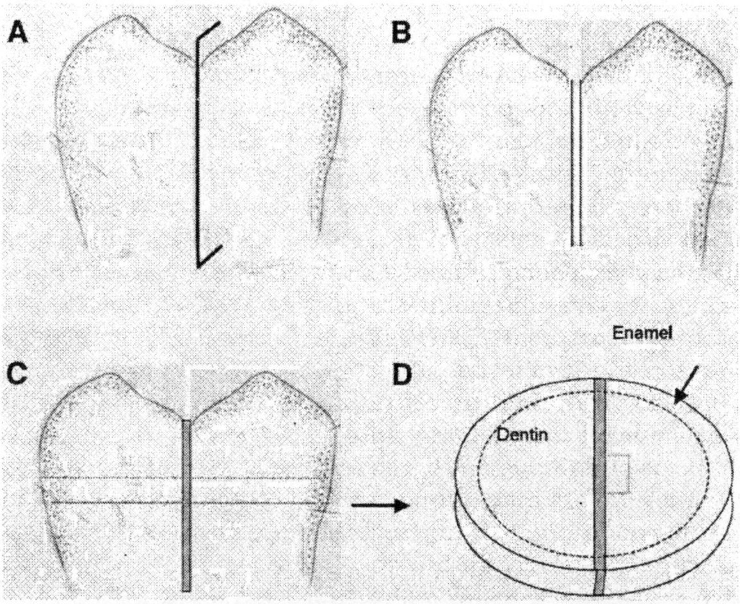

Fig. 3. Schematic diagram for construction of the cyanoacrylate reference-layer method. (A) Sample is sagittally sectioned; (B) two halves are polished; (C) two halves are bonded together and a disk is cut parallel to the occlusal surface; (D) the disk has an enamel periphery and inner dentin region with a thin layer (not drawn to scale) of cyanoacrylate that is used for the reference layer and area for study (small square).

6. After removal (Fig. 3D), the disk has a periphery of enamel surrounding the den-tin and a thin cyanoacrylate bonded layer of about 10 μm in thickness through its center. This layer serves as the height reference layer in studies of demineraliza-tion and solution effects on the dentin.
7. The surface to be studied (either top or bottom) is then prepared using the same polishing steps, with ultrasonic cleaning between steps, which were described previously for the cut surfaces that were bonded together to reassemble

the tooth. Various areas to one side or the other of the reference layer can be studied, first at baseline in water, and then after exposure to the selected treatment solution for various times.

Solutions

Typically, a dilute acid solution is prepared from reagent grade chemicals. Solutions with a pH of about 2 work well with this method and allow sequential measurements of the changes in height, relative to the reference layer, to be made at different locations in each field of view. However, nearly any solution of interest, including commercial etchants, can be used. Concentrated solutions etch the structure more rapidly and therefore the structural changes may be difficult to follow, which is the reason that dilute solutions are preferred. For example, in dilute solutions at pH 2–3, the changes in peritubular dentin can be sequentially measured after 5-s exposure steps for 20- to 60-s cumulative exposure time. After this, it is difficult to see the peritubular dentin because it etches more quickly and recedes below the level of the surrounding intertu-bular dentin. In contrast, a single 5-s exposure with a more concentrated acid, for example, 10% citric acid or 35% phosphoric acid, typical of those used for bonding procedures, will etch the peritubular dentin below the surface and will not allow contact with the AFM tip, so measurements cannot be made. Measurements of the changes in the intertubular dentin can be made with dilute or concentrated solutions.

Imaging and Demineralization Treatments

1. Usually initial images (baseline, *see* Fig. 4A) are taken by placing the polished dentin disk in the wet cell of the AFM and taking three to four images along the reference layer with dimensions of 20 μm × 20 μm to 50 μm × 50 μm (*see* Note 2).
2. Height differences between the reference layer and various locations in the den-tin then can be determined, using the section analysis software program (Fig. 4). The image is processed using plane fit analysis procedures on the reference layer that was highly polished and is assumed to be flat. Site-to-site measurements are made of height differences between the reference layer and various

locations on the peritubular dentin surrounding the dentin tubules and the intertubular dentin areas between the tubule units (Fig. 4B).

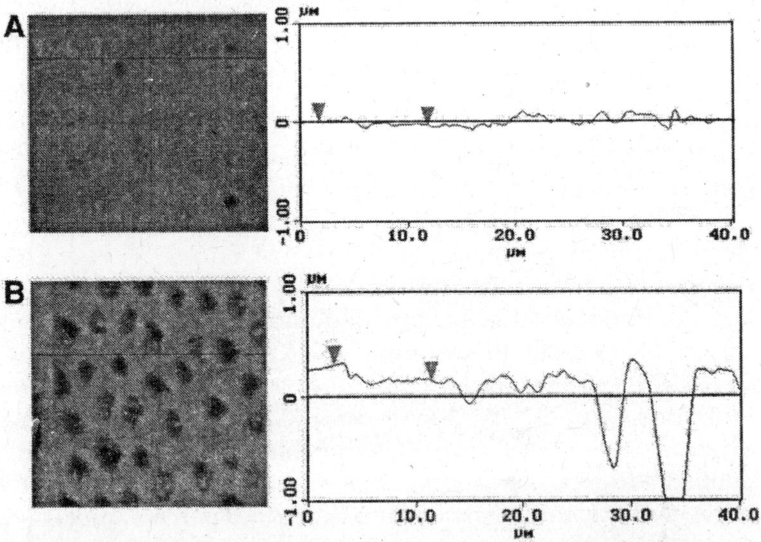

Fig. 4. Example of baseline and etched samples of dentin demineralized with citric acid These images are from the same area and the reference layer is at the left. (A) Baseline image in water of a 40 × 40 µm field of view. Many tubules are filled because this image was taken from transparent dentin that has mineral deposits inside tubule lumens. At the right side of the image is a section analysis showing height differences between selected points on the reference layer and the intertubular dentin (difference was 41 nm). (B) Same area after 30 s of etching showing enlarged tubule lumens and recession of the intertubular dentin as shown by the shading differences and the section analysis at the right. The height difference between selected points was 153 nm.

3. Normally, three to five measurements each for peritubular dentin and intertubu-lar dentin are made per image. If there are other features of interest, such as intratubule mineral, these can be measured in the same way *(20,21)*.
4. After baseline imaging, the disk is removed from the AFM and the desired solution is applied for a selected time. Application of the solution can be done in a number of

ways, including immersion or application with a small sponge that is commonly provided in bonding kits (Sun Medical Co., Ltd, Moriyama, Japan).

5. After uniform application of the solution for the desired time the sample is thoroughly washed in purified water and placed back in the AFM wet cell and each of the previously imaged areas is reimaged.
6. Each image is plane fit, and measurements are made between the reference layer and the previously selected points (Fig. 4B). For dilute solutions, we usually use 5-s exposure increments for times up to 30- to 60-s cumulative exposure. The peritubular dentin is etched rapidly during this time period and gradually recedes below the surrounding intertubular dentin, leaving enlarged tubule lumens (Fig. 4B). Exposure increments are continued in steps of 10 s to a minute or more to follow continued changes in the dentin structure and the recession characteristics of the intertubular dentin. Normally we perform this procedure for cumulative periods of 1800 s, although shorter and much longer times have been used, depending on the aims of the study. The intertubular dentin also recedes initially at a high rate, but more slowly than the peritubular dentin. The intertubular recession generally slows, and after some time, appears to change very little with additional etching. Thus, when recession is plotted vs time for the intertubular dentin the surface recession appears to slow or reach a plateau. This is attributed to replacement of the mineral in the intertubular dentin with water that compensates for the differences in volume. However, if the dentin is dehydrated at any point during the procedure, the partially demineralized dentin will rapidly collapse. Thus, any dehydration will lead to errors in the measurements of surface recession and should be avoided. This is an additional motivation to make all measurements in a wet cell and under water.

It should be noted that brief drying can be reversed rapidly, and therefore, inadvertent drying during handling will be overcome by re-immersion in the wet cell.

Sequential Surface Recession, Etching Rates, and Etching Characteristic Curves

For a selected field of view in a baseline image the point-to-point differences in height between the reference layer and each of the selected peritubular dentin locations and intertubular dentin areas give the initial surface height. Ideally, if all areas are perfectly polished, the differences would be zero, but in practice there is usually a small difference because the peritubular dentin, intertubular dentin, and reference layer; all have different hardness and different response to polishing. Our experience has been that peritubular dentin protrudes slightly above the intertubular dentin and the reference layer is usually at the same or a slightly higher level as well (Fig. 4A). The same locations are measured for each exposure interval using plane-fit processed images, based on the assumption that the reference layer is flat. Clearly, the more accurately the same points can be selected in sequential images, the more accurate will be the measurements of recession.

Figure 4B illustrates a 30-s etch in dilute citric acid for the area shown in Fig. 4A and illustrates the measurements of height difference between the peritubular dentin and the reference layer, and between the intertubular dentin and the reference layer using the section analysis procedure. After all the height differences from all the images are collected and corrected for any difference in height between the selected points and the reference layer at baseline, recession-vs-time curves can be constructed for peritubular dentin and intertubular dentin. As noted earlier, the recession curve for peritubular dentin can be measured only for a short time because of the rapid recession below the intertubular dentin that does not allow accurate contact with the pyramid-shape of the AFM tip. However, with data at exposure times between 0 and 20–30 s, the recession-vs-time curve can be used to estimate the etching rate of the peritubular dentin for the given solution. A similar procedure can be used for intertubular or other dentin structural components. However, the slowing of recession and the appearance of an apparent plateau in the recession for long periods makes it difficult to characterize the intertubu-lar dentin behavior from a single short-term rate.

We usually estimate the level and/or time at which the plateau occurs to characterize the intertubular dentin etching behavior (*see* Fig. 5). If this procedure is adopted, the features of the plateau are used rather than an estimated etching rate to describe the intertubu-lar dentin. This seems appropriate because the AFM tip measures the intertu-bular dentin surface location, which is really a demineralized collagen network after an etching exposure, rather than the location of the demineralization front. The demineralization front increases with depth of continued etching although the surface recession appears to plateau, i.e., the recession rate approaches zero while the true etching rate does not. The difference between the location of the demineralization front and the apparent plateau can be determined with other techniques such as dehydration of the specimen, high-resolution computed tomography imaging, or scanning electron microscopy of fractured cross-sections.

The apparent plateau is probably not a real plateau and in fact depends on the overall depth of etching. When the location of an apparent plateau is compared with the extent of etching, samples with markedly deeper total deminer-alization demonstrate higher values for the level of the apparent plateau (greater recession). Thus, we speculate that the maximum value for the apparent plateau depends on total demineralization depth. This value could be determined by total demineralization of the sample and would, therefore, depend on the thickness of the original dentin disk. The experiment implied by this speculation has not as yet been conducted. If true, this means that the recession of the intertubular dentin would slowly increase with increased demineralization of the sample until a true plateau is reached, which is dependent on the thickness of the dentin disk. Despite these limitations, the apparent plateau and the time to reach this value has proven valuable in characterizing differences in dem-ineralization response with different solutions and with different types of dentin, such as caries-affected transparent dentin and sclerotic dentin. However, this also raises the question of how best to characterize the intertubular recession curve, define the plateau, and compare recession curves for intertubular dentin treated in different experiments. We have used different approaches to this problem. The apparent level of the plateau can be defined by a fixed criterion, for example, the point at which there is a change in dimension of less than some amount, from then until the conclusion of the experiment. With this level defined, statistical comparisons can be made to compare either the time or level of the plateau for different experimental conditions.

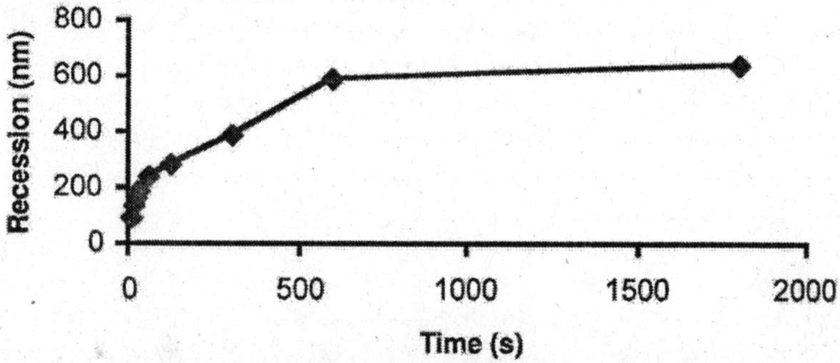

Fig. 5. Long-term decreases in height relative to the reference layer (recession) for demineralized intertubular dentin as a function of etching time in dilute citric acid for normal dentin. The recession level or time at which a plateau is observed is often useful in describing the characteristics of a particular form of intertubular dentin.

Other Dimensional Change Measurements

The effect of dehydration leading to the collapse of the demineralized dentin matrix (collagen) was previously mentioned and is of considerable interest from the standpoint of dentin bonding procedures, clinical treatments, and understanding the role of moisture in the dentin structure. In some experiments we have evaluated the effects of dehydration and rehydration using both the preceding reference layer method and other masking methods. A difficulty arises in using the cyanoacrylate method because the etching procedure weakens the interface between the reference layer material and the dentin. On dehydration, the drying stresses often disrupt the layer and therefore do not allow measurements to be made. The weakening of the interface, as might be expected, increases with the extent of demineralization, so that for samples with shallow demineralization, the effect of dehydration and its reversal by rehydration can be followed. However, our experience has been that with deeper (prolonged or higher concentration solutions) the interface is largely destroyed and other methods need to be used.

AFM-Based Nanoindentation for Hardness and Elasticity Measurements of Dentin

The same sample preparation methods can be used to prepare samples for mechanical properties measurements using the AFM. Although initial measurements were reported using a stiffer cantilever and diamond tip, improved equipment allowing recording of load-displacement curves now permits both hardness and elastic modulus determinations to be made. For these kinds of measurements the standard AFM head from the Nanoscope III is replaced with a Triboscope indenter system. In this configuration, the standard AFM head is replaced by a capacitive sensor. The sensor consists of two fixed outer drive plates that are driven by AC signals 180° out of phase relative to each other. Because of the small spacing between the two plates, the electric field changes linearly from one to the other.

Therefore, the electric field potential is highest at the drive plates and zero at the center between the two plates. The center, or pickup, electrode is suspended in a manner so that it moves up and down in the region between the two drive plates. The pickup electrode assumes the electric potential of the space between the two drive plates. This results in a bipolar output signal that is equal in magnitude to the input signal at the maximum deflection, and zero at the center position. The synchronous detector converts the phase and amplitude information from the sensor output into a bipolar DC output signal. The output signal is actually a reading of the pickup electrode position. In the imaging mode, this signal is used as a feedback to the piezoceramic tube for constant force contact imaging. In the indentation mode, the feedback is cut off and a voltage ramp is applied to the lower drive plate. As a result, an electrostatic force is generated between the pickup electrode and the drive plate. The force can be described as follows:

$$F = k_e V^2 \tag{1}$$

where k_e is the electrostatic force constant and V is the applied voltage. The voltage ramps are formulated to produce triangular, trapezoidal, or square force loading profiles of the sample. For experiments in air, the force is applied to the sample through a diamond tip glued to a tapped polymer holder attached to the pickup electrode by a small screw. In liquid, the tip is glued to a tungsten rod with a large aspect ratio, which in turn is attached to the polymer holder. In this configuration, the diamond tip and portion of the tungsten rod are immersed in the solution and as a result the meniscus force remains constant as the height of liquid changes because of vaporization.

This force can be easily accounted for before contacting the sample. In the imaging mode the minimum contact force applied to the sample is about 1 | iN. In the indentation mode, loads up to 30 mN can be applied. A variety of indenter shapes are available, including Berkovich and cube corner indenter. Because image quality depends on the tip radius-of-curvature, the image quality using this system may not be as good as that using the conventional AFM cantilever. However, imaging with the indenter tip allows precise selection of the area or region to be indented, and postindent imaging allows verification of the location and sharpness of each indent. The sharp cube corner indenter displaces more than three times the volume of the Berkovich indenter at the same load.

The cube tip is particularly useful when attempts are to be made to fracture a particular feature. The hardness, *H*, is calculated on the basis of maximum force, F_{max}, divided by the projected contact area at maximum load, *a*, whereas the indention or reduced elastic modulus, *E**, is calculated from the contact stiffness, *S*, defined as the slope of the linear portion of the force/displacement curve during unloading near the maximum load. The stiffness, *S*, can be determined from a polynomial expression for the unloading curve that is obtained from a least-squares fit of the data between 20 and 95% of the maximum force, and then the derivative of the unloading curve evaluated at the peak force is used as the contact stiffness, *S*:

$$H = F_{max}/a \quad (2)$$

*E** = /i/ffi · £ (3)

V *a* 2

These equations appear to be straightforward, but several steps need to be taken to obtain correct values. The projected contact area, *a*, is usually expressed in terms of contact depth, *h*. For an ideal pyramidal geometry tip (e.g., Berkovich), the relationship is simply as follows:

$$a = C0h^2 \quad (4)$$

For a real tip the relationship can be conveniently expressed as follows:

$$a = C0h^2 + C1h + C2I1^{1/2} + C_3h^{1/4} + C\wedge^{1}/^{8} + C5I1^{1/16} \quad (5)$$

where C_0 for a Berkovich tip is 24.5, whereas for a cube corner tip it is 2.598. To determine the relationship between h and a for a specific tip, a series of indents at various contact depths are performed on fused silica specimens and the contact area, a, is calculated using Eq. 5. Finally, a plot of the computed area as a function of contact depth is plotted and fitted to Eq. 5 to obtain the coefficients in this equation.

The Young's modulus of the probed specimen, E_s, can then be obtained from E^* and the known modulus of the diamond indenter, E_i, and the known or assumed values for Poisson's ratios, í:

$$1/E^* = (1 - í_s^2)/E_s + (1 - í_i^2)/E_i \quad (6)$$

Indentations can be made at intervals of 1–2 µm using indents of submicro-scopic size (approx 300–500 nm). The spacing is needed to avoid influence of one indent on the adjacent indent. Using this approach indents can be made on individual structural components of the dentin or across junctions, such as the dentin–enamel junction, as shown in Fig. 6.

AFM-Based Nanoindentation of Demineralized Dentin

There is considerable interest in measuring dentin or other calcified tissues as modified by demineralization or hypomineralization. For example, etching procedures are used widely for bonding dentin; dentin caries (tooth decay) introduces modifications of structure that have various levels of mineralization; and genetic anomalies, such as dentinogenesis imperfecta, result in dentin of reduced mineral level. In other calcified tissues, hypomineralized or hypermineralized tissue states are of great interest. A detailed discussion of the methods for study of such calcified tissue alterations are beyond the scope of this work, but several factors need to be mentioned. Balooch et al. used a modified indenter (flat-ended cylinder) to evaluate the viscoelastic properties and low elastic modulus of fully demineralized dentin when hydrated (approx 0.1 GPa) and dried (2 GPa), whereas normal hydrated intertubular mineralized dentin has an indentation modulus of about 20 GPa. Enamel modulus values are less than 100 GPa, so that AFM-based indentation methods can be used for measurement of calcified tissues over a range of at least three orders of magnitude.

However, the measurements of demineralized tissue or partially dem-ineralized tissue become extremely complex. Sharp indenter tips cannot be used reliably when the substrate consists of a demineralized feltwork of collagen fibrils.

When a portion of the sample is demineralized and an underlying area remains mineralized, it becomes uncertain whether the underlying mineralized portion of the sample affects the measurement. We recently have studied several of these complex substrates.

It is unlikely that indentations of areas with indentation modulus values below 2 GPa can be made with conventional tips. However, the limit is difficult to establish and we currently rely on careful examination of the indent shape. Furthermore, calibration of the indentation equipment and determination of the indentation tip properties are

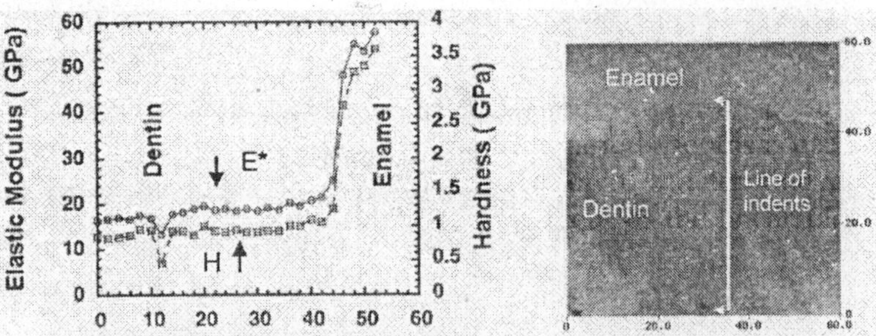

Fig. 6. Nanoindentation of enamel, dentin, and the junction between them (DEJ). The image at the right shows the line of indentations. The variations in hardness and indentation modulus as a function of position for each of the indentations are shown in the graph based on Eqs. 2 and 3.

currently carried out using quartz standards. This is an excellent procedure for mineralized tissues, but there is no adequate soft standard to insure that reliable indents can be made of the highly demineralized structures.

Notes

1. The cyanoacrylate viscosity is important. If the viscosity is too low, it will penetrate into the dentin and fill the tubules. This becomes evident during deminer-alization as the filled tubules become increasingly obvious as surrounding dentin recedes leaving them sticking up above the surface. Several cyanoacrylate gels have the desirable viscosity but were not found to be stable. Thus each product should be checked in the solution of interest. The MDS QX4 ethyl a-cyanoacry-late that we now use is too thin as dispensed and we therefore air dry a drop of the material for 2 h before bonding the polished halves of the teeth.
2. Placing fiduciary marks on the opposite side of the dentin that is to be studied can be used to facilitate identification of the same areas at each step of the demineral-ization experiment.
3. To obtain the real value of contact stiffness, *S*, we need to account for compliance (the inverse of stiffness) of the machine, C_m *(8,10)*. The measured compliance is the sum of machine compliance C_m, and the contact compliance, C_c:

Therefore, a linear plot of $C_{measured}$ and $1/4a$ for indentations made at various depths is used on a silica standard (for which C_m is well known) to determine machine compliance, which is simply the intercept of the linear plot on the compliance axis.

12

Applying Atomic Force Microscopy to Studies in Cardiac Physiology

INTRODUCTION

At the present time there exists a great deal of interest in the application of scanning probe microscopy methods to the imaging of cellular systems. It would now not be an exaggeration to state that atomic force microscopy (AFM), in particular, represents perhaps the most powerful means of structural/functional analysis at the level of a single live cell. In recent years this technology has been applied, amongst many other systems, to studies of bacterial flagella, erythrocytes, human platelets, endothelial cells, skin fibroblasts, plant cuticles, and cardiac myocytes. In resolution terms, perhaps the most impressive work is that of Engel et al.. Although not on whole cells (thereby greatly simplifying the experiment), molecular-level images of isolated cellular gap junctions, which play an important role in intracellular communication and signal transduction, have been obtained. More recently, experimental protocols have advanced to the point where it is possible to monitor, simultaneously, both cellular topography and ion channel flux. The ability to characterize functionally active cellular systems at a nanometre resolution under controlled fluid conditions has also been used in the monitoring of time-dependent cellular change.

Importantly, it is now generally accepted that, under appropriate imaging conditions, nanometre-resolved AFM experiments can be conducted with negligible levels of cellular damage or, indeed, impaired cell physiological function. Despite this, however, under normal operational forces, one generally resolves submembrane structure rather than the structure of the membrane itself though the tip is only in direct contact with the latter. The plasma membrane offers little resistance to the loading force exerted by the scanning tip and is accordingly deformed as the tip traces out underlying internal structure. Although this surface indentation has hindered high-resolution imaging of the extracellular surface, it can be used in the spatial characterization of internal cellular components. In summary, scanning probe methods, especially when combined with optical systems (confocal, total internal reflection, epifluorescence) provide a very powerful means of resolving the three-dimensional structure and dynamics of cellular systems under ambient and near-physiological conditions.

Cell Immobilization/Pretreatment

For any biological structure to be imaged satisfactorily by scanning probe methods, it must be immobile on an underlying substrate surface, under the typical lateral forces imparted by the scanning probe, at least for the duration of the scan. A range of methods has been developed for cellular immobilization. Most simply, the cells are dried down onto, or grown on, a suitable flat substrate surface; commonly mica (an aluminosilicate) or glass microscope slides. Nonspecific or electrostatic coatings, such as laminin or poly-L-lysine, are commonly applied to the bare surface to enhance the immobilization stability. Perhaps most simply, cells can be grown directly onto a substrate surface; these typically adhere robustly, although their structures are commonly perturbed (they may appear, for example, to be somewhat "spread" on the surface—a common observation with strong cell–substrate adhesion).

Cells have also been entrapped in flexible, porous media or imaged while embedded in a resin. It can be useful to adjust cellular mechanical properties (particularly in terms of membrane flexibility) by fixation (such as with glutaraldehyde). Such cell pretreatment generally enhances image resolution of the cytoskeleton. Related to this, the advantages associated with lowering imaging temperature, in terms of improved cellular robustness, have been noted.

Basic AFM Requirements

The substrate should have a surface roughness well below the vertical dimensions of the cell (and typically to be flat on a nanometre scale). In both contact and tapping force imaging it is commonly useful, if not critical, to quantify the forces imparted by the scanning tip (usually for the purposes of minimizing structural deformation of the sample). This is achieved, once the spring constant is determined (or estimated), through the generation of force– distance or amplitude–distance curves. The process of acquiring force–distance data over a cell surface can indeed be used to map cell mechanical properties with high spatial resolution. Note also that, under solution, it can be instructive to consider the effects of pH and ionic strength in minimizing electrostatic contributions to imaging force. Tip-sample adhesion forces are also highly dependent on the chemistry at the probe surface (be this silicon, silicon nitride, or gold) and tip cleaning in oxidizing acids, or oxygen plasmas can be an effective way of attaining a reproducible behavior.

Cardiac Myocytes

The human heart is shaped like a blunt cone and is the size of a clenched fist, being some 12 cm long and 9 cm wide on average and weighs between 250 to 390 g in an adult male and about 200 to 250 g in a female. The mammalian heart is a hollow organ, containing four chambers, with a wall of muscle (or septum) dividing the organ vertically into a right and left heart. At the top of each half is an atrium (porch or antichamber) whereas below on each side is a ventricle (little belly). Within the walls of the heart are cells that both conduct electrical impulses and contract intermittently to raise blood pressure in the atria and ventricles. Particularly within the ventricular walls, these myocytes are arranged in bands as either spirals (like a turban) or in horizontal layers. The overall geometry of this contractile machinery subserves the function of impelling the blood from the left ventricle, or squeezing it out of the right half of the heart, into either the systemic or pulmonary circulations. The left ventricular wall is made up primarily of powerful circular musculature, sandwiched between layers of spiral muscles running from base to apex. The right ventricular wall consists almost exclusively of spiral muscles. This intricate myo-cardial anatomy, coupled with the distinct morphology and function of specialized component cells, often presents acute problems in the interpretation of experimental data obtained from the whole organ. Over the last 25 yr, individual myocytes isolated from the adult heart have become an important model system for a vast range of investigations concerning cardiac function in both the normal and diseased states. We shall demonstrate here that these myocytes are proving very useful in structure-function studies at resolutions obtainable with AFM.

Materials

1. High-purity buffer reagents; iV-hydroxyethylpiperazine-iV-2-ethanesulfonate (HEPES), Dulbecco's minimal essential medium (DMEM).
2. Phosphate-buffered saline (PBS); (Sigma-Aldrich, cat. no P 0261).
3. High-purity water; 18.2 MQ cm (Elga, UK).
4. 0.2 um Syringe filters (Sterelin, UK).
5. Cleavable mica substrate (Agar Scientific, UK).
6. Glass microscope slides (Agar Scientific, UK).
7. Optical microscope (preferably of the Inverted configuration, with 60 to 100x objectives).
8. Scanning probe microscope capable of ambient and fluid contact and noncontact imaging.
9. Etched silicon and silicon nitride probes of appropriate spring constant and resonance frequency.
10. Perfusion apparatus; standard system for retrograde coronary perfusion (*see* ref. *24* for details).
11. DMEM solution (Life Technologies, UK).
12. Serum substitute; Ultroser G (Life Technologies, UK, cat. no. 81-003).
13. Collagenase; Worthington Type CLS I (Lorne Laboratories, UK).
14. Glutaraldehyde (Sigma-Aldrich).
15. 10-mL Sterile tubes (Sigma-Aldrich, cat. no. C-3084).

Centrifuge Note: with heart cells, small benchtop centrifuges are not to be recommended because their rapid acceleration can damage myocytes. Much better are the larger free-standing cooled types, such as the Mistral 4L.

METHODS

The methods described below outline the process of (1) cell preparation, isolation, and electron microscopy characterization and (2) scanning force characterization.

Cell Preparation and Electron Microscopy Characterization

The detailed tissue dissociation protocols for the adult heart will provide isolated myocytes suspended in Dulbecco's modified Eagle's medium supplemented with pyruvate and 10 m*M* HEPES to which has been added 2% (v/v) serum substitute. It is convenient to store cells at room temperature in sterile 10-mL tubes from Sigma-Aldrich. As long as standard precautions have been taken during isolation, these myocytes can be used for up to 24 h without any antibiotics.

When incubation at 37°C is used (100% O_2 with HEPES media or 95 % O_2/5% CO_2 v/v if bicarbonate buffered only) then antibiotics must be added. Myocytes can be plated directly on to cover slips. Alternatively, cells can be fixed by first resuspending in Krebs' buffer (NaCl, 134.1; KCl, 5.4; $NaH_2PO_4.2H_2O$, 0.3; $MgCl_2$, 1; Na acetate.$3H_2O$, 5; Na pyruvate, 5; glucose, 11.1; HEPES, 5; all m*M*, pH adjusted to 7.4 with NaOH) and then adding 1 volume of myocytes to 2.5 % v/v glutaraldehyde in 1/3 Krebs' buffer. Though fixation is rapid, the cells are typically left for at least 30 min, then resus-pended in phosphate-buffered saline and stored at 4°C. This fixative procedure has been designed to reduce the osmotic strength of the fixation solution and thereby reduce cell shrinkage.

High yields of ventricular myocytes from adult mammalian hearts are obtainable using retrograde (Langendorff) coronary perfusion of the whole organ, with fluids low in calcium, containing collagenase and protease. Details of the procedures have been presented elsewhere. Examination of suspensions of isolated myocytes by bright-field light microscopy showed two distinct cell types; a dominant rod-shaped form together with a population of rounded cells (Fig. 1). It is clear from a number of tests (dye exclusion, morphology from low-power light microscopy and electron microscopy) that the rounded cells are myocytes damaged by the dissociation protocols. Standard purification procedures can reduce the round cell population to <10 % of the total. Of major interest are the rod-shaped cells, which display many of the gross morphological characteristics expected from studies of whole cardiac tissue.

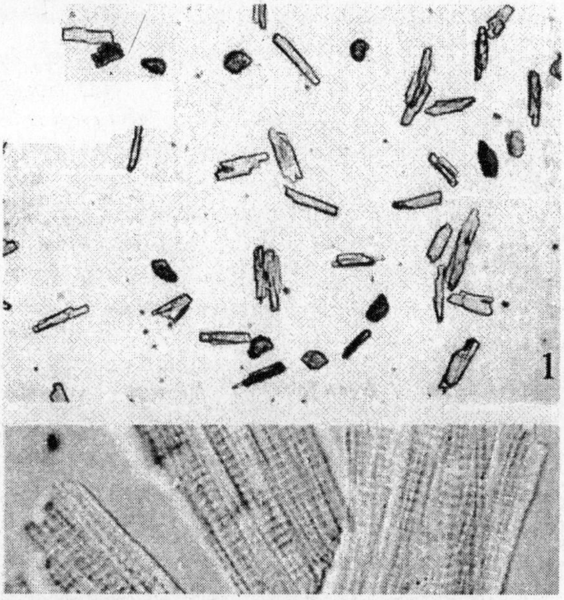

Fig. 1. (1) Bright-field survey view of a preparation of isolated ventricular myocytes. Damaged cells appear dark and round and are easily distinguished from the intact rod-shaped cells. (magnification ×100). (2) Light microscope view at higher magnification (×800) showing the structural details of rod-shaped cells. (3) Electron micrograph of a longitudinal thin-section through an isolated rod-shaped myocyte. Characteristic banded-structure myofilament bundles alternate between rows of mitochondria. In this example, there are 64 sarcomeres along the length of the cell. (magnification ×1500).

Isolated myocytes are more rectangular than cylindrical with irregular profiles, and also large, being some 80–180 μm long, 8–20 μm wide, and 8–16 μm thick. This wide range of sizes reflects the complex ventricular ultrastructure, which is illustrated more clearly by scanning electron microscopy, where individual cells are seen to be most irregular in shape (Fig. 2) with longitudinal grooves. On occasion, very flat cells are also observed (Fig. 3A). If it were possible to fit the cells back together again, then it becomes clear how the spiraling and horizontal bands of whole muscle (see above) arise from the complex interdigitation of myocytes with grossly varying shapes. Also apparent in Fig. 2 are the regular transverse ridges on the surface of each cell. These reflect the action of the underlying contractile proteins, which run longitudinally within each myocyte and partially overlap, giving rise to the striated appearance of cardiac muscle, when longitudinal sections are viewed with the electron microscope (Fig. 1, Plate 3). As will be seen below, these general morphological features of cardiac ventricular myocytes are apparent at magnifications comfortably achievable by AFM.

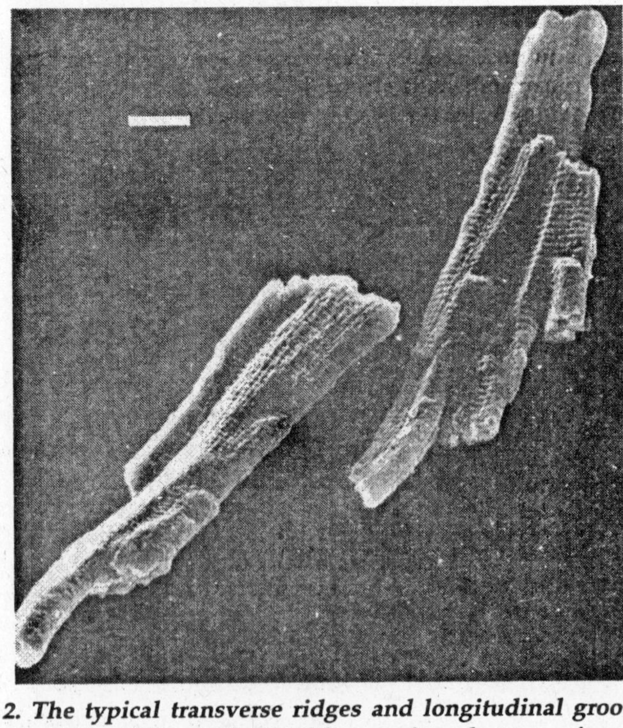

Fig. 2. The typical transverse ridges and longitudinal grooves of heart muscle cells, obtained with scanning electron microscopy. Scale bar, 10 μm (magnification ×1158). From ref. 26, with permission from Elsevier.

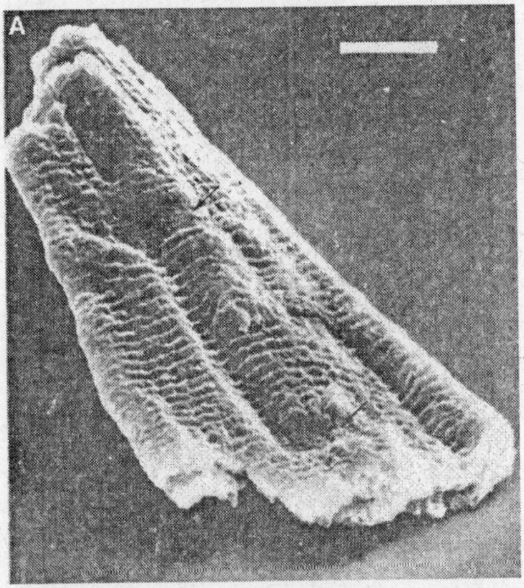

Fig. 3. (A) Scanning electron micrograph of cell having flatter profile than those shown in Fig. 2. Scale bar, 10 µm (magnification ×1158). From ref. 26, with permission from Elsevier.

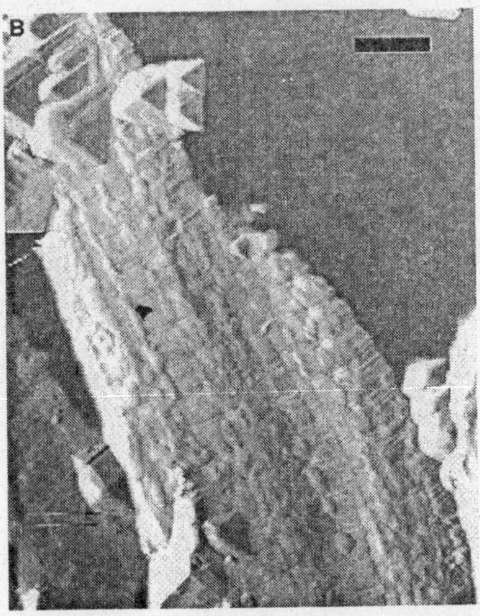

Fig. 3. (B) Large-scale deflection AFM micrograph of a single isolated cardiac myocyte immobilized on mica substrate. Note the intercalated disc region evident on the right-hand-side of the cell. Scale bar, 5 µm. From ref. 9, with permission from Elsevier.

The scalloped appearance of the surface membrane is seen in more detail in Fig. 4A. Evident at these levels of magnification are the regular apertures in the surface sarcolemma, the mouths of tubules dipping transversely into the cell. These T-tubules conduct the wave of electrical depolarization traveling across the myocyte surface down into the cell, so that contraction is triggered synchronously throughout the cell interior during each heartbeat. These scan- ning electron micrographs, taken on air-dried specimens of fixed myocytes, are directly relevant to observations made by AFM (see paragraphs following and Note 2).

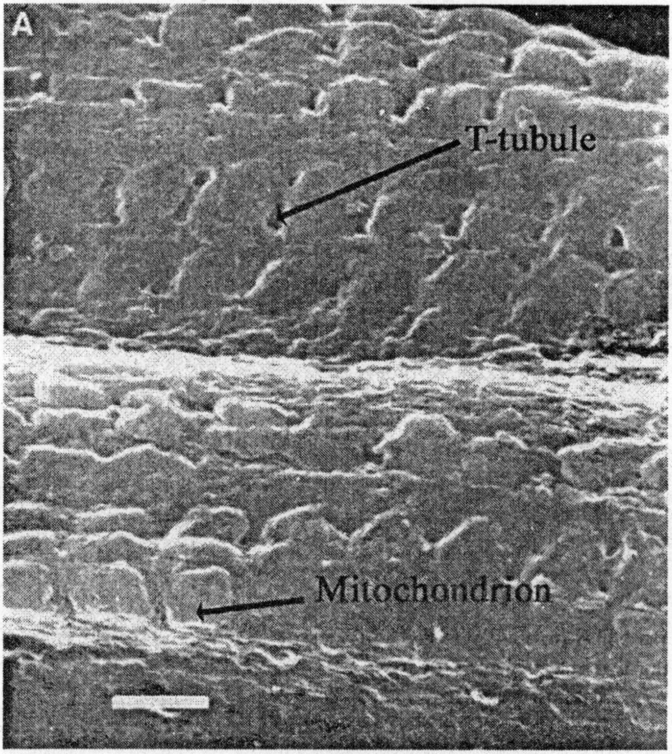

Fig. 4. (A) Scanning micrograph showing array of T-tubule openings, which have both ovoid and circular cross-sectional geometry. Scale bar, 2 µm (magnification ×8000). From ref. 26, with permission from Elsevier.

Under these conditions, the surface membrane is forced down on the underlying structures, and in Fig. 4A there are seen rectangular "packets" at the periphery of the cell, lying along what appears to be a cylindrical structure within the cell. These packets are mitochondria, the oxidative motors in the heart, providing the aerobic production of ATP for the contractile proteins, which is essential for a normal mechanical output from each ventricle in the whole organ. It follows that if scanning force microscopy is applied to myocytes prepared using the same protocols, similar subsurface structures should be apparent.

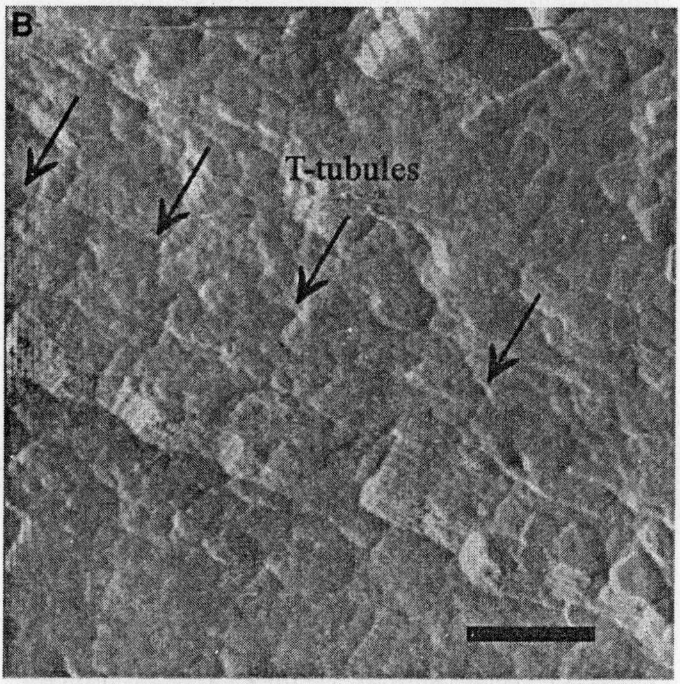

Fig. 4. (B) Deflection AFM micrograph of a myocyte surface in which T-tubule openings are indicated by arrows. Compressed between the contractile machinery are mitochondria. Scale bar, 2 µm.

Figure 5 demonstrates that all the normal features of sarcolemmal structure, including the glycocalyx, are preserved in these isolated cells. The structure of intracellular membranes and organelles also shows excellent preservation, indistinguishable from that reported in cells of the intact heart. A network of tubules in the M region, the M rete (MR), is connected by longitudinal elements (L) to a z tubule (ZT_L) seen here in longitudinal section.

Examples of transversely sectioned z tubules, seen elsewhere, are indicated by ZT_T. Junc-tional sarcoplasmic reticulum (JSR) is continuous with the free SR (arrow) and consists of flattened cisternae closely apposed to the sarcolemma. Peripheral junctional SR (JSR_p) occurs in association with the surface sarcolemma, and interior junctional SR (JSR_I) occurs against transverse-tubule membrane. The latter may take the form of dyads (D), which consist of a transverse tubule plus one JSR cisterna, or triads (T), which consist of a transverse tubule sandwiched between two JSR cisternae. The JSR lumen is characteristically bisected by an electron-dense line, and regularly spaced electron-dense feet project from the membrane facing the sarcolemma. Mitochondria (M) show well-preserved membranes and cristae.

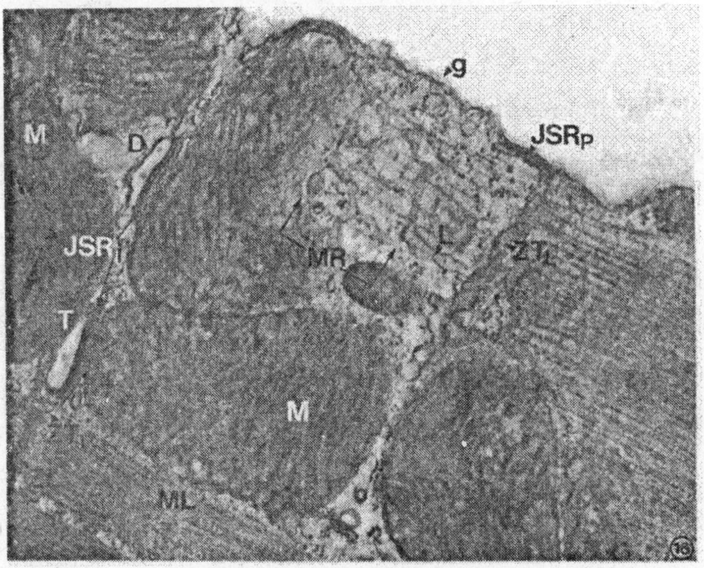

Fig. 5. Electron micrograph detailing the structural complexity of the interior of mammalian ventricular heart cells. See text for further description and abbreviations. (magnification ×50 000). D, dyad; M, mitochondria; JSR, junctional sarcoplasmic reticulum; JSR_P, peripheral JSR; L, longitudinal elements; ZT_T, transversely sectioned Z tubules; ZL, Z line; ZT_L, longitudinally sectioned Z tubule; ML, M line; g, glycalyx. From ref. 26, with permission from Elsevier.

Finally, if the plasma membrane is freeze-fractured, intramembrane particles can be observed (Fig. 6) reflecting aggregations of membrane proteins involved in cellular function; the density of these particles in isolated myocytes is very similar to that observed in whole hearts (Fig. 7). Note that the scale bar for these measurements (100 nm) is approaching the range more usually associated with scanning force microscopy. Clearly, freeze-fractured membranes would make most interesting specimens for AFM imaging. From the evidence presented here, and from many other studies, of those myocytes surviving the dissociation procedures, the structure of intracellular membranes and organelles shows excellent preservation, indistinguishable from that reported in cells of the intact heart.

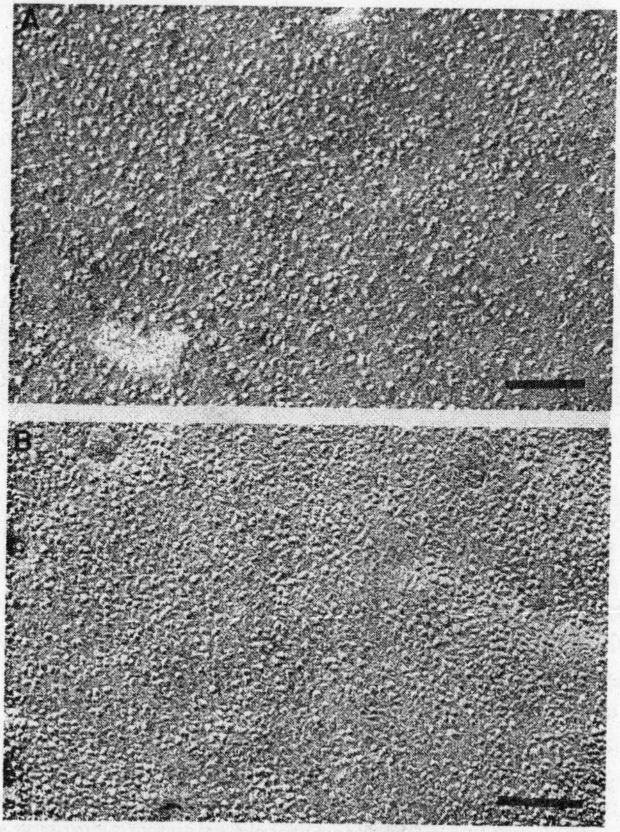

Fig. 6. (A) P-face view of isolated myocyte sarcolemma. (B) P-face view of myocyte sarcolemma from intact left-ventricular myocardium. Scale bars 0.1 µm.

AFM

Aliquots of myocytes (approx 100 μL of cell suspension) were added to a freshly cleaved mica (Agar Scientific, Cambridge, UK) surface. The sample was then gently blown dry with high purity argon or allowed to dry overnight at 5°C. Experiments were performed with a Digital Instruments MultiMode microscope in conjunction with a Nanoscope IIIa control system. A "J" scanner (with a lateral range of approximately 125 l im) was used. Etched silicon probes, attached to triangular cantilevers 100-200 l im in length, were used (Digital Instruments, UK Ltd, model TESP) operated at resonances in the 300-400 KHz range (nominal force constant 20-100 N/m).

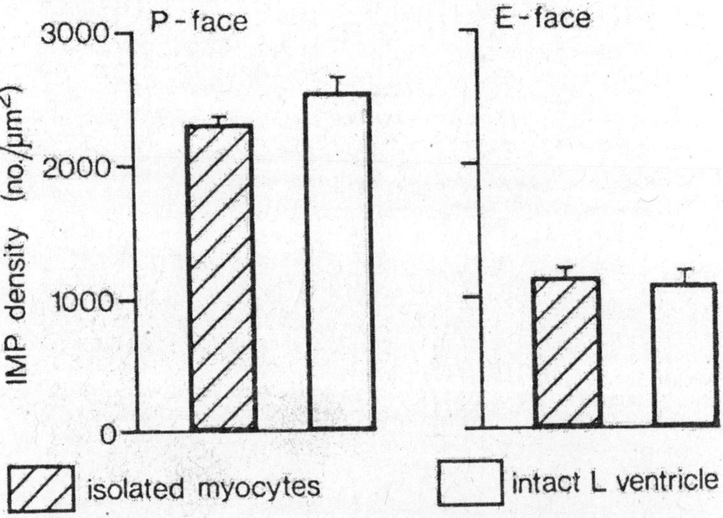

Fig. 7. Numerical density of membrane particles (expressed as numbers per square micrometer) for isolated myocytes and intact myocardium. From ref. 27, with permission from Elsevier.

Drive amplitude was adjusted so as to give the sharpest resolution at the minimum amplitude (typically 1-2 V rms). Integral and proportional gains were balanced so as to allow simultaneous acquisition of sharp, noise-free, height and amplitude data sets. With the aid of a x30 magnification eyepiece, the scanning cantilever was positioned directly above a surface-immobilized cell. Because the vertical dimensions of the cells exceed the full vertical range of the scanner (ca. 4 l im), scanning laterally into a cell commonly led to destruction of the probe. Vibrational/acoustic shielding was achieved by mounting the microscope in a PicoIsolation chamber (Molecular Imaging Co) during scanning. Height, amplitude and phase data were simultaneously collected, the latter with a Digital Instruments Phase Extender Module. Data sets were subject to a first order flattening and low band pass filtering only when stated. Thermal noise levels were estimated to be approx 0.4 Å.

As shown in Fig. 3, whole-cell images obtained by AFM methods compare very favorably with scanning electron micrographs. Even at this relatively low magnification, both the cellular cross-striations and the step-like morphology of the intercalated disc regions are clearly evident. Fine details of the subsurface structure were, in general, more easily observed in deflection mode than height mode, though imaging was somewhat sensitive to the scanning set point and drive amplitude (equivalent to energy dissipation by the probe rather than the imaging force directly). Scanning at high amplitude produced no noticeable structural damage (though increased deformation of the plasma membrane is likely as the drive amplitude is increased), that is, when the same area was subsequently reimaged at lower drive no significant image deterioration was observed. The fact that cell dimensions exceeded the full vertical range of the scanners prevented a quantification of the effect of increased dissipation on cell height though changes in surface roughness were minimal across the range of values used. At increased magnification, the general morphology of working ventricular myocytes can be seen in more detail (Fig. 8). This AFM image strikingly reveals all the major characteristics of cellular morphology that have been discussed previously in reference to electron microscopy (*see* Note 6). The parallel arrays of longitudinal contractile machinery, separated by embedded organelles and transversed by regular tubular structures to demarcate sarcomeres, all reflect the integrated structure to be expected of this syncytial tissue.

The scalloped nature of the external sarcolemma, with the grooved surface structure reflecting the underlying contractile apparatus alternating between rows of mitochondria, is shown in Fig. 4B, which is presented with a corresponding scanning electron micrograph to emphasize the remarkable similarity in the two images. Z grooves, which run at right angles to the long axis of the myofilaments, mark the sarcomeres from one Z line to the next and can be quite deep and narrow, especially in contracted tissue. It is clear that there is a consistent relationship between these grooves and the Z lines, suggesting strongly that the Z line material must be attached firmly to the interior face of the plasmalemma. Using such images, resting sarcomere length is measured as 1.6–2 μm and T-tubules of diameter 200–260 nm are present in rows at approximately every 1.8–1.9 μm. The measured lateral spacing between T-tubule openings (Fig. 4B) is 2–2.3 μm. These dimensions are comparable to those reported for fixed ventricular cells.

Notes

1. It is clear that to exploit the many advantages that AFM imaging has to offer for the study of heart cells, the obvious prerequisite is a preparation of stable, isolated myocytes. Though more than 30 yr have elapsed since dissociation techniques were first reported, this initial step remains one of the most difficult. Much has been written about isolation protocols (*see* references in 24) and here we can give only a few of the main stumbling blocks that might prevent successful dissociation of heart tissue. The first very important procedure is retrograde coronary perfusion of the whole heart. A simple alternative would be to chop the organ and incubate the tissue chunks in the appropriate solutions, but this produces very low yields of poor quality myocytes. Cardiac tissue is predominantly aerobic, so any delay in excising and mounting the organ for in vitro perfusion can cause irreversible damage to the constituent cells. Experience has shown that the highest-quality reagents should be used in the perfusion fluids (but not generally of spectroscopic grade) made up with high quality distilled water. Ideally, the whole dissociation protocol should be conducted in a sterile environment and the cells stored under tissue culture conditions, but as long as standard levels of cleanliness are observed, this should not be a limiting condition. Like most physiological investigations, living tissue is very susceptible to minor changes in buffer quality; therefore, new reagents should be screened before routine use. We, for example, purchase

enzymes and serum albumin in bulk to minimize any major variations in myocyte quality.

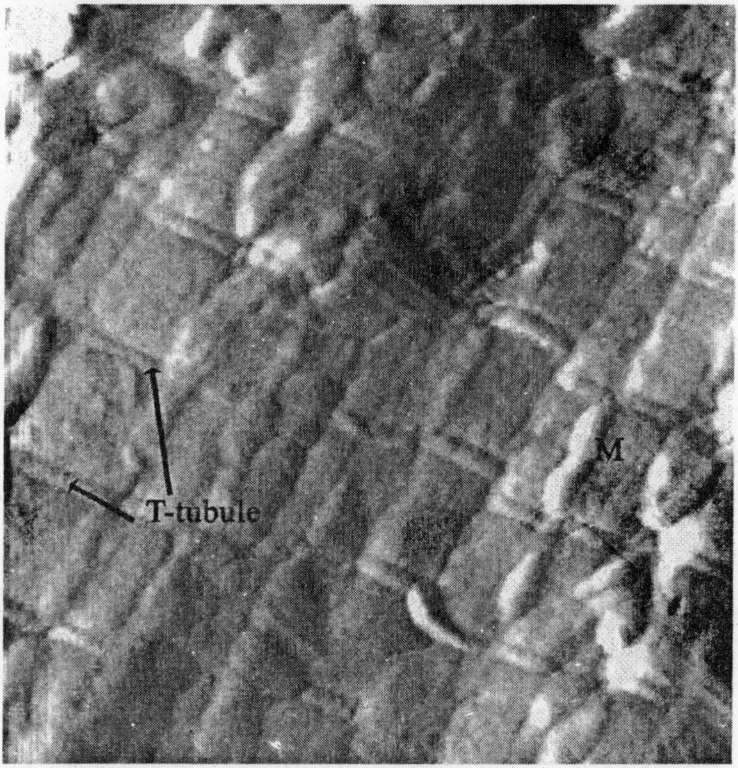

Fig. 8. High-resolution deflection AFM micrograph of a myocyte cell surface. Longitudinal columns of underlying contractile proteins, separated by valleys having regular transverse grooves. The globular features evident (M and elsewhere) are subsurface mitochondria. Also evident are the invaginations of the surface membrane by transverse tubules. Scale bar, 2 µm. From ref. 9, with permission from Elsevier.

2. The cells of choice should be immobilized on a flat, stable, and (ideally) optically transparent substrate surface. This may require pretreatment of cell and/or substrate surface.
3. The scanning force microscope is set up with a low spring, sharp lever and allowed to stabilize. High-powered inverted optics are used to locate the cells and carefully position the scanning probe above the one of choice. This is of particular importance with very large cells, such as

cardiac myocytes, as they are typically about 8 μm thick. Tapping Probes, in particular, are expensive and easily damaged during alignment or if scanning is initiated when the tip lies to the side of a cell.

4. Surface-dried isolated myocytes are easily visible in the bright field. The cells become opaque and considerably more difficult to see in bright field on immersion in fluid. Higher optical magnification (>500×) under such conditions is of considerable benefit.
5. It is important that the cells be mechanically stable under the lateral forces imparted by the scanning probe. The worst-case scenario is when the cell is dragged across the supporting surface as scanning starts. Several imaging modes should, ideally, be available, for example, contact, noncontact, "tapping," magnetic AC ("MAC"). Each has its advantaged and disadvantages and selection should be based on maximizing resolution at minimal cellular perturbation. Occasionally, and especially in fluid, imaging will suddenly deteriorate as matter adsorbs to the scanning tip. Though commonly terminal (in terms of the tips usefulness), it may be possible to clean the probe using oxygen plasma.
6. Although surprisingly underused by physiologists and cell biologists, it is clear that AFM imaging techniques can be adapted for a large variety of cellular studies and, in terms of three-dimensional topographic resolution, compete well with classic scanning electron microscopy. By integrating AFM technology with commercially available (or adaptable) optical microscopes (such as the Nikon TE2000 inverted) one can generate a powerful imaging and characterization system. Such multiport systems can incorporate, rather straightforwardly, epi-fluorescence and confocal microscopy (with electronic motorized transfer between the two) and do not require complex set-up and alignment procedures. By integrating characterization systems in this way, it should be possible to address important questions of structure–function relationships in living cellular systems or preparations.

13

Imaging Bacterial Shape, Surface and Appendages

INTRODUCTION

Bacteria are typically smaller than eukaryotic cells. The average diameter of *Staphylococcus aureus* is 1 ± 0.5 | im, whereas *Escherichia coli* is on average 0.5 x 1.5 | im. The bacterial cell is also characterized by the presence of a complex external rigid structure called cell wall, which protects the internal protoplast and gives also the cellular shape, that generally falls into one of the, three basic morphologic categories, spherical (cocci), rod-shaped (bacilli), and spiral. Some bacteria show an atypical bacterial shape.

Bacteria are also able to extrude some material that collects outside the cell wall to form an additional, surface layer. Many genera of bacteria possess also filamentous structures projecting through the cell wall to form the so-called surface appendages. The most commonly observed bacterial appendages are flagella, fimbriae or pili, and filaments.

Antibiotics are particular type of drugs able to interfere in different ways to the metabolic pathways of bacteria. This causes also changes directly or indirectly in the structure of cell wall and consequent alterations in the shape of bacteria. The integrity of cell wall and bacterial shape are important to maintain the vitality and the virulence of bacteria. Morphostructural alterations not only cause bacteria to loose cytoplasm but also to be more easily phagocytized and killed by human phagocytic cells. A large amount of basic and clinical researches in microbiology, chemotherapy, and infectious diseases have been performed to investigate the morphology and structure of bacteria. These studies have been previously conducted by means of optical microscopy and scanning electron microscopy (SEM).

Optical and scanning (or transmission) electron microscopes are classified as far-field microscopes because the distance between the sample and the point at which the image is obtained is long in comparison with the wavelengths of the photons or electrons involved.

In this case, the image is a diffraction pattern and its resolution is wavelength limited: in optical microscopy, resolution is determined by the Nyquist relation to the wavelength of the light used (typically about 1 μm); in a general purpose SEM, it is limited by the properties of the electromagnetic lenses (typically about 50 Å).

In 1986, a completely new type of microscopy was proposed: without lenses, photons, or electrons, it involved the mechanical scanning of samples *(4)* and opened up unexpected possibilities for the surface analysis of biological specimens. Initially called the scanning force microscope (SFM), it was a development he previous scanning tunneling microscope *(5)*, which provided information at atomic resolution of specimens that are electrically conducing.

Because SFMs involve interactions between atomic forces (about 10^{-9} Newton), they are also and more frequently called atomic force microscopes (AFMs).

These new types of scanning probe microscopes (SPMs) are based on the concept of near-field microscopy, which overcomes the problem of the limited diffraction-related resolution inherent in conventional microscopes. Located in the immediate vicinity of the sample itself (usually within a few nanometers), the probe records the intensity and not the interference signal, and this greatly improves resolution.

As shown in Fig. 1, AFM explores the surface of a sample not by means of a system of lenses that form an image using the diffraction patterns of rays of different wavelengths but by means of a very small sharp-tipped probe located at the free end of a cantilever driven by the interatomic repulsive or attractive forces (van der Waals forces) between the molecules at the probe tip and those on the surface of the specimen. This can be done by scanning the sample laterally (x, y) while a closed loop control system keeps the tip in proximity to the surface by adjusting the z position of the sample.

In most AFMs, tip movements are monitored by reflecting a laser beam from the back of the cantilever on to a position-sensitive photodiode.

AFM is extremely useful for analyzing the three-dimensional structure of the surface of biological specimens, particularly bacteria. Although SEM is still frequently used, the introduction of the AFM technique offers substantial benefits in real quantitative data acquisition in three dimensions, minimal sample preparation times, flexibility in ambient operating conditions (i.e., no vacuum is necessary), and effective three-dimensional magnification at the submicron level.

To investigate the shape and the surface of bacteria offers the possibility of investigating the efficacy and the mechanism of action of antibiotics that disrupt this structure as an epiphenomenon of internal biochemical action *(9–13)* and at the same time the possibility of investigating their lack of activity, as in the case of resistance.

MATERIALS

1. Test organisms. Both Gram-positive and Gram-negative bacteria are suitable for AFM.
2. Tryptic soy broth or other suitable medium.
3. Phosphate buffer saline (PBS): 0.02 *M* phosphate and 0.15 *M* NaCl, pH 7.3.
4. Glutaraldehyde: 2.5%.in 0.1 *M* cacodylate buffer, pH 7.0.
5. Graded alcohols (60, 70, 80, 90, and 100%).
6. Incubator.
7. Centrifuge.
8. Micropipette and sterilized disposables for culturing bacteria.
9. Round glass coverslides, diameter 6–7 mm (or mica).
10. AFM (including probe-tips, software for processing signals and three-dimensional rendering, and computer).

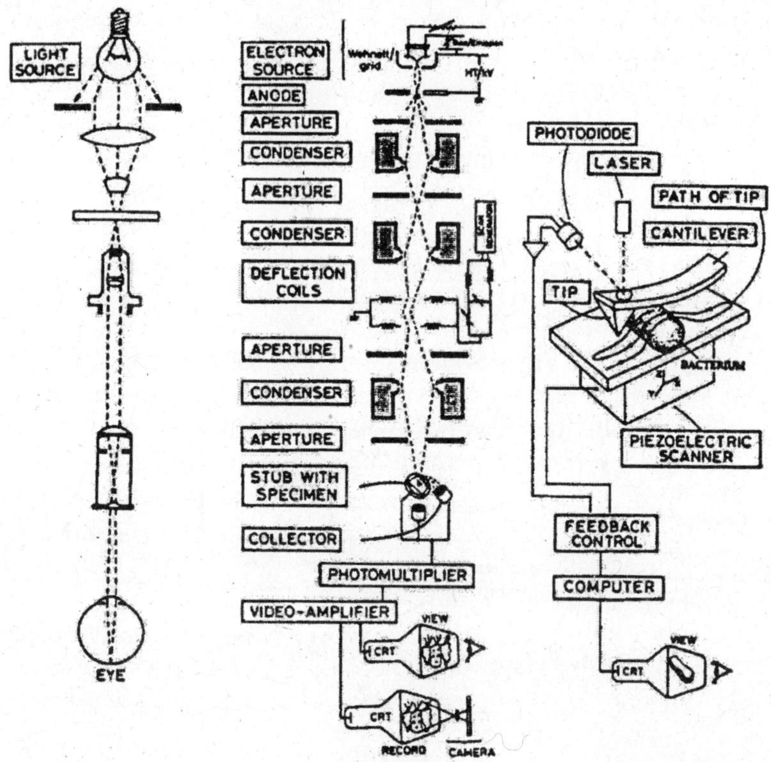

	LIGHT MICROSCOPE	SCANNING ELECTRON MICROSCOPE	ATOMIC FORCE MICROSCOPE
-SAMPLE			
-preparation	none/ little	freeze drying gold sputtering vaccuum	none/ little
-characteristics:	thin/ trasparent	vacuum compatible	not excessive variation in height
-operating environment	dry/liquid/ vacuum	vacuum	dry/liquid/ vacuum
-BIOLOGICAL APPLICATION	many	surface analysis	surface analysis
-FOCUS (depth)	medium	small	small
-FIELD (depth)	small	large	medium
-RESOLUTION			
xy axis	1.0 μm	5 nm	0.1-1.0 nm
z axis	——	——	0.01 nm
-MAGNIFICATION	$1\text{-}2 \times 10^3$	10×10^4	$5 \times 10^2\text{-}10^8$

Fig. 1. Comparative schematic view of the elements characterizing light microscopy, scanning electron microscopy, and atomic force microscopy, together with their specific technical parameters.

METHODS

1. Prepare the cultures of chosen microorganism according to common standard procedures.
2. Wash the test microorganism from the suspension in broth (i.e., 10^6 cells/mL) three times with PBS.
3. Resuspend the final pellet in 1–2 mL of PBS.
4. Collect 0.1 mL (or less) of this suspended bacteria with a micropipet and place it on round glass coverslide.
5. Dry the coverslip in air.
6. Fix with 2.5% glutaraldehyde in 0.1 *M* cacodylate buffer (pH 7.1).
7. Dehydrate in graded alcohols.
8. Dry the coverslip in air.
9. Repeat steps 1–8, incubating bacteria with various supra-minimum inhibitory concentrations (MICs) or sub-MICs of antibiotic.
10. AFM observation. A typical AFM imaging session begins by firmly fixing the sample cover slide to the microscope holder to avoid even the slightest movement and then positioning it under the probe tip and locating the area of interest by moving the x–y table.
11. A good-quality on-axis optical microscope is essential to be able to position the probe tip in the proximity of a bacterium to be imaged by AFM. Because bacteria are about 1 μm in size, it is necessary to have appropriate lighting conditions to distinguish them from any debris on the slide surface. In the experiments described here, a reflection optical microscope equipped with long-range objectives was used. Although the cantilever bearing the probe partially obstructs the optical view of the underlying bacteria, it does allow the probe to be positioned sufficiently accurately in the area of interest. Commercial AFM instrumentation coupled to a transmitted light optical microscope offers a higher degree of precision in the first approach of the probe to the sample.
12. Once an area has been located after the tip-to-sample approach, a first large scan (i.e., 30 × 30 μm) using a high scan speed and small number of pixels per line can be

made in order to assess its exact position within the scanner coordinate system, identify the nature of the bacteria and select an interesting one. Further smaller scans may be necessary in order to position the bacterium exactly at the centre of the scanning area.

13. Record high-resolution images by using appropriate instrument settings depending on the imaging mode selected (contact, intermittent contact, non-contact). In general, accurate feedback setting is necessary to obtain the maximum possible gain for the resolution of bacterial surface structures while avoiding oscillation when scanning along the cell sidewalls.
14. Acquire image (typically acquired at 512 × 512 pixels) and process by means of plane fitting, high-frequency filtering, and three-dimensional-shaded rendering (Figs 2 and 3).
15. Postprocessing analysis and the spatial representation of AFM-generated data are essential to extract all of the available information from the image dataset. Because the recorded data are an intrinsically three-dimensional digital matrix (the height of the sample recorded at each x, y coordinate), the software makes it easy to obtain numerical data of cross-sections of interesting features expressed with sub-nanometer accuracy. The same software allows three-dimensional rendering of the surface and rotation in space so that only one acquisition is needed to be able to observe the same object from many different points of view.

Notes

1. It is better to use low concentrations of bacteria because they tend to concentrate in small areas during the air-drying phase, whereas a single bacterium provides a clearer image. Be sure to mark the location of your specimen on the upper surface of your round glass coverslide to avoid wasting time investigating the wrong side.
2. If the sample is kept dry, repeated sessions could generally be performed without any loss of resolution.
3. Bacterial sample preparation for AFM is very simple and rapid. There is no need for critical point drying, which also

avoids shrinkage effects; there is no need for gold sputtering, a procedure that covers and smooths fine surface details. There is no need for vacuum conditions, as with SEM.

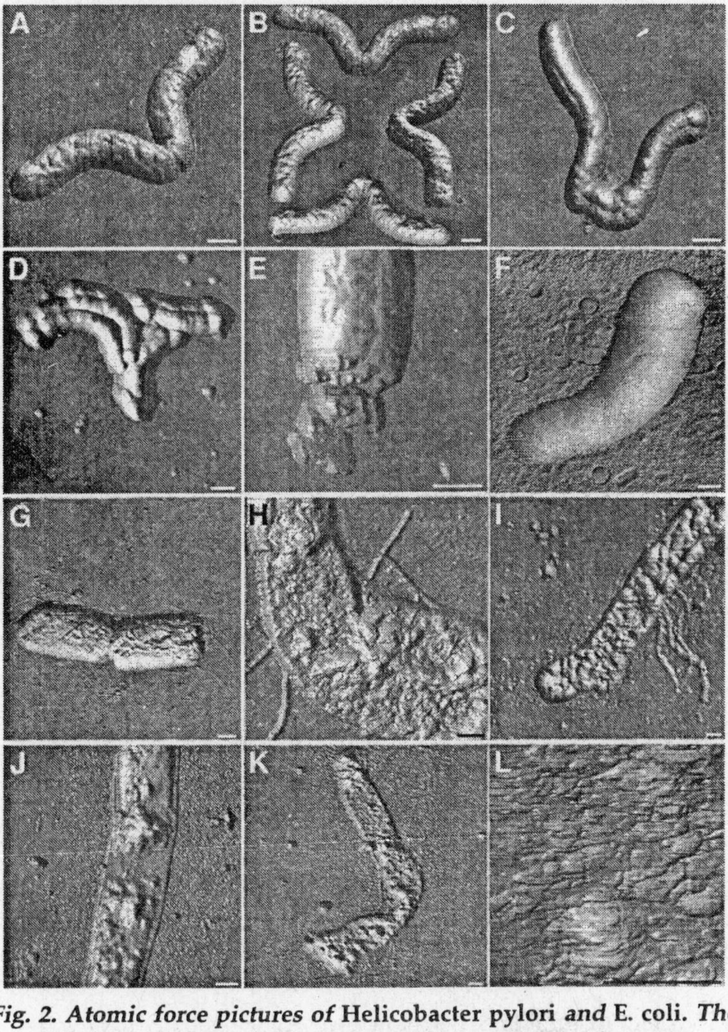

Fig. 2. Atomic force pictures of Helicobacter pylori *and* E. coli. *The bar at the bottom right of each picture corresponds to 500 nanometers. (A) Common morphology of* H. pylori *without exposure to antibiotic. (B) Example of the different perspectives obtained by means of computer processing (C–E) Different alterations induced in* H. pylori *by*

4. A recent technical evolution has also opened up the possibility of using AFM on wet samples, that is, living cells immersed in biological fluids in culture chambers.
5. Care must be taken when choosing the adhesive used for fixing the glass slide to the sample holder. Avoid using thick double-sided adhesive tape, as this can expand for a long time after pressure and thus cause instability in the vertical position of the tip. The specially produced sticky tabs made by different manufacturers are fine.
6. To obtain the best results, it is necessary to be thoroughly familiar with the characteristics of different cantilevers and tips, how these can be used and how they suit to the different kinds of samples investigated. For high-resolution work, tip sharpness is essential: tip properties can vary significantly within the same batch of cantilevers. Fine tuning of the feedback loop and set point, together with the chosen scan speed, is critical for good surface tracking.
7. As mentioned above, AFM offers different imaging modes for investigating the sample. There is the contact mode in which the tip of the probe makes soft physical contact with the sample, which should be used with harder and stiffer materials than biological samples as it can easily give rise to undesirable effects as the result of tip-to-sample interactions (Fig. 2L). Tip pressure can indent and deform the sample surface, and lateral forces can stretch the sample, drag away loosely bound fragments, or even detach the whole bacterium from the substrate. These drawbacks of the contact AFM mode are overcome by using the intermittent-contact mode, also called tapping mode. In this case, the AFM feedback loop constantly dampens the high-frequency oscillations of the vibrating cantilever due to the tip coming into contact with the surface for a very short time. For this reason, indentation effects are less invasive, lateral forces are greatly reduced, and a high lateral resolution can be maintained. In the third noncontact mode, small amplitude and high-frequency oscillations induced on the cantilever allow the feedback control loop to maintain the tip-to-sample distance within the range of attractive Van der Waals forces.

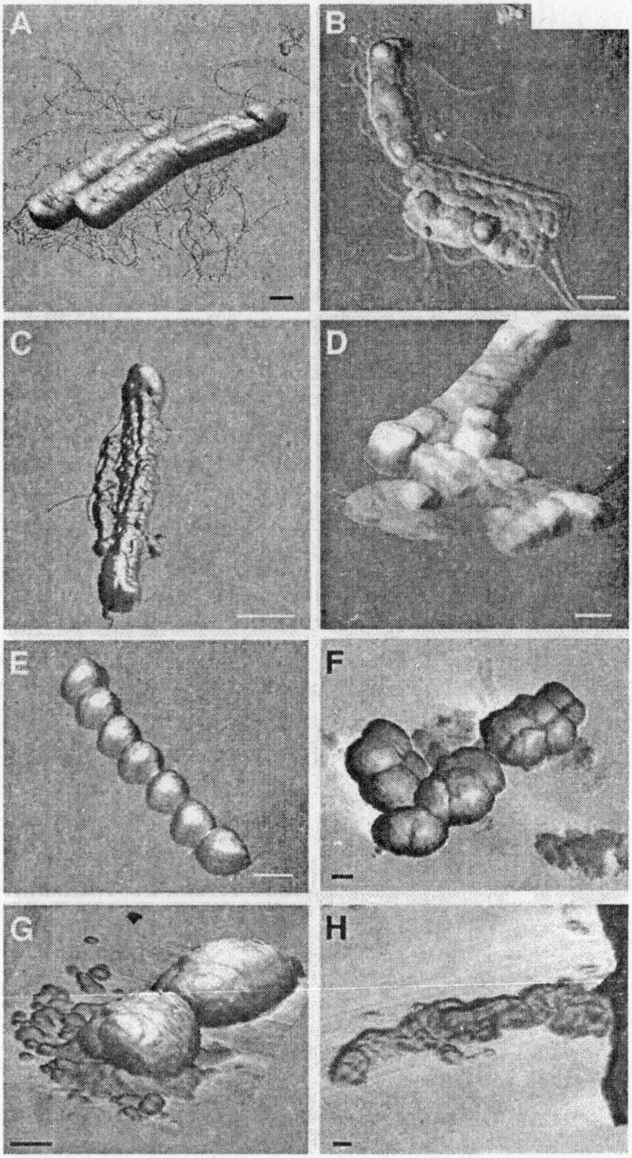

Fig. 3. Atomic force pictures of Bacillus cereus *and* Streptococcus pyogenes*. The bar at the bottom right of each picture corresponds to 1 μm. (A) Example of untreated common rod-shaped morphology with flagella of* B. cereus*. (B) Morphostructural*

Tip-to-sample interactions are greatly reduced at the expense of lateral resolution and the scanning speed. For biological specimens the noncontact and intermittent contact are the most suitable, although the contact mode may be used for high-resolution work on very small areas.

8. To make accurate dimensional measurements, the calibration of the AFMs piezoelectric scanner has to be periodically checked. The procedures are usually described in the instrument manual. Lateral dimension calibration is relatively straightforward, but special care must be taken when calibrating height. We used a VLSI standard calibration grid (NIST traceable) with a 100-nanometer nominal step height and an in-house developed statistical analysis procedure for calibration.
9. The images may sometimes be blurred as a result of poor washing procedures, an electrostatic charge on the specimen, improper feedback parameter settings, debris on the tip, or an eroded tip (Fig. 2F).
10. The images of spherical bacteria, such as *S. aureus*, will suffer from little lateral resolution along the perimeter because of the perpendicular direction of analysis. In general, the shape of the tip and its lateral walls will limit the detection of steep elevated features (Fig. 4).
11. After image acquisition, the built-in software allows the rendering of the picture to be greatly improved by means of shadowing, rotation, different illumination, and different points of view (Fig. 2B).

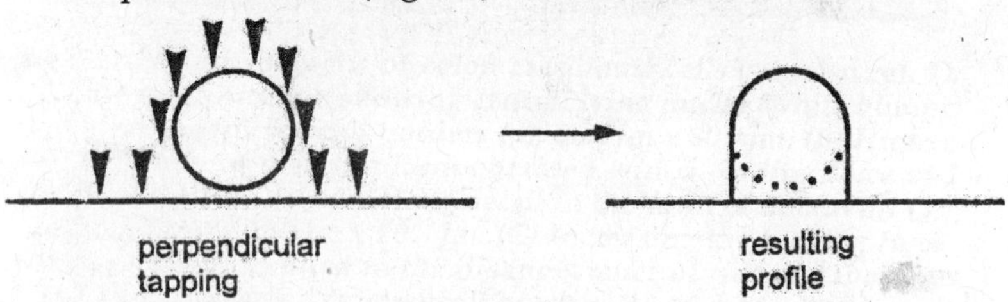

Fig. 4. Example of the AFM rendering of a spherical bacterium.

(14)

Visualizing Nuclear Structure in Situ by Atomic Force Microscopy

INTRODUCTION

The cell nucleus is an organelle where molecules involved in gene expression are highly compartmentalized in very dynamic territories. This current notion of functional organization has been possible because of many studies of this organelle that included its cell and molecular organization, and where microscopy played an important role. Light and electron microscopy in conjunction with molecular approaches, such as the construction of both antibodies and nucleic acid probes, as well as molecular construction of genes with green fluorescent protein and fluorescence recovery after photobleaching technology, in addition to high-resolution *in situ* hybridization and immunocytochemistry now offer a more complete knowledge.

Structurally, the mammalian cell nucleus is organized as an organelle surrounded by a double layer of phospholipids called the nuclear envelope, which is interrupted by the nuclear pores. It is covered in the interior by the nuclear lamina. Within the cell nucleus, a nuclear matrix is present as a nuclear skeleton made of several different proteins. DNA is organized as euchromatin and heterochromatin. The ribonucleoproteins are distinguished by their participation.in pre-mRNA or pre-rRNA metabolism. In the first case, ribonucleopro-teins are visualized as nuclear speckles by light microscopy, which correspond to interchromatin granule clusters and perichromatin fibers observed by electron microscopy. In addition, perichromatin granules are also present. All these structures are related to transcription and splicing of pre-mRNA and transport or storage of intranuclear mRNA. The nucleolus, however, is involved in pre-rRNA transcription and processing as well as ribosome assembly.

There are other nuclear structures, such as the Cajal bodies, which contain factors involved in both pre-mRNA and pre-rRNA metabolism. In plants, there are two major classes of nuclei: (1) the chromocentric and (2) the reticulated types, depending on the arrangement of heterochromatin fibers. In the first type, DNA is organized as discrete clumps within the nucleoplasm. In the second one, DNA forms a reticulated strand.

Lacandonia schismatica is a rare plant whose most important feature is the inverted position of the sexual organs, with the androecium in the center, surrounded by the gynoecium. In studying the biology of this species, we have analyzed the cell biology, including the interphase cell nucleus, which we have published previously. The cell nucleus of *L. schismatica* is reticulated.

To explore the possibility of studying the interphase cell nucleus with high resolution, while at the same time of working in solution, we have been studying the nuclear structure by using an approach whereby a sample is prepared for transmission electron microscopy (TEM) and the surface of the unstained semithin sections is explored with an atomic force microscopy working in contact mode to visualize the interior of the cell.

It is a common observation when sectioning with an ultramicrotome that trimming of the plastic blocks to get thin sections many times reveals the profile of the embedded cells over the surface. This observation suggested that even very flat surfaces have a texture corresponding to cell organelles, and indeed this is the case.

Therefore, because the microscope is an instrument to analyze surfaces, the surface of each section can be visualized as a representative sample of a portion of the cell interior. Similar approaches also have been used previously *(21,22)* to study cell structure *in situ*. As a first step, we have been validating this approach by observing and generating images of already known material.

MATERIALS

Sample Preparation

1. 70% Glutaraldehyde (Polysciences). Toxic. Wear gloves, goggles and work in a chemical fume hood. Store at 4°C.
2. 2% Osmium tetroxide. Toxic. Wear gloves, goggles and work in a chemical fume hood. Store at 4°C. Protect from light.
3. Phosphate-buffered saline (PBS) 0.01 *M*, pH 7.4. Store at 4°C.
4. Deionized water.
5. 100% Ethanol, electron microscopy grade.
6. 100% Propylene oxide, electron microscopy grade.
7. Epoxy resin (glycidether 100, Merck). Plastic resin.
8. DDSA (dodecenylsuccinic anhydride) hardener (Merck).
9. NMA (methylnorbornene-2,3-dicarboxylic anhydride) plasticizer (Merck)..
10. DMP (2, 4, 6-Tris[dimethyl-aminomethyl]phenol) accelerator (Merck)..
11. Toluidine blue.
12. Clean glass slides. Keep within a chamber to avoid dust.
13. Ultramicrotome (Reicher-Jung).
14. Glass knives.
15. Diamond knife.

Microscope

1. BioScope (Digital Instruments, Santa Barbara, CA).
2. NanoScope IIIa (Digital Instruments, Santa Barbara, CA) control system and the software.
3. Inverted Diaphot 200 microscope (Nikon, NY).
4. Silicone nitride tips, 20–60 nm radius of curvature (model NP).
5. 100-μm Atomic force microscope scanner.
6. Proscan software (version 3.1, Park Scientific, 1997)

METHODS

Sample Preparation

Samples are prepared as for standard TEM or histology. For TEM, we have used several protocols, including different types of fixation with aldehydes such as glutaraldehyde or paraformaldehyde at different concentrations, or with embedding media such as glycol methacrylate, London resin white, Lowicryl®, and epoxy resin. Here, we describe the regular protocol for TEM *(15,23–25)* because it produced thus far the best results as far as morphology is concerned.

1. Samples are fixed with freshly made 6% glutaraldehyde from 70% vials, buffered with 0.01 *M* PBS at pH 7.4 for 2–6 h at room temperature. PBS is prepared as follows *(23)*: dissolve 8 g of NaCl, 0.2 g of KCl, 1.44 g of Na_2HPO_4, and 0.24 g of KH_2PO_4 in 800 mL of distilled water. Adjust pH to 7.4 with HCl. Add water to 1 L. Sterilize by autoclaving.
2. After fixation, samples are washed with PBS for three times 10 min each.
3. Postfixation is performed with 1% osmium tetroxide in bidistilled water for 1–2 h. Prepare from a stock solution of 2% made from crystals in bidistilled water.
4. Dehydration is conducted with a series of graded concentration of ethanol, 10 min each of 30, 50, 70, 80, 90, and 96, and three changes of 100% ethanol. Pro-pylene oxide is then used for three times for 10 min each.
5. For preembedding, samples are placed in a mixture of 1:1 of propylene oxide and epoxy resin during 16 h at room temperature. The epoxy resin is made as a stock solution. A recommended proportion of the components for working with plants can be taken from the references *(23)* and will contain 9.44 g of epoxy resin, 5 g of DDSA, 10 g of NMA, and 0.2 mL of DMP.

1. Mix the components thoroughly with a glass rod avoiding making bubbles. Store at 4°C in a small jar avoiding the penetration of air by sealing with parafilm. Prepare in the chemical fume hood. Avoid contact with the skin and to inhale.

6. Embedding is conducted with epoxy resin for 16 h at 60°C.
7. Semithin sections of about 150–250 nm width are obtained with an ultramicro-tome working with glass or diamond knives.
8. Two sets of sections are placed onto an ethanol-cleaned glass slide. The sections are transferred by placing a platinum loop onto a drop of water. The sections are placed onto a glass slide. Sections are fixed to the glass by gently heating the slide. One set of sections is stained with toluidine blue and rinsed thoroughly with bidistilled water. Toluidine blue stain is prepared as a stock solution by dissolving 1% toluidine blue in 1% sodium borate, to stain the sections but not the plastic. For staining, once the slide reaches room temperature after fixing to the glass, cover the section with one or two drops of the stain and place it on a hot plate until stain starts to dry and obtain a metallic green color at the borders. Rinse the excess stain off the slide with bidistilled water. Sections are stored in a chamber to avoid dust.

Atomic Force Microscopy

An atomic force microscope model Bioscope (Digital Instruments) is used for observations. The microscope is mounted on an inverted light microscope Diaphot 200 (Nikon). Observations are made in contact mode. The scan size are from 100 to 5 µm at a scan rate varying from 1.969 to 1.285 Hz. Images were generated with the NanoScope IIIa control system. Alternatively, an atomic force microscope from Park Scientific has been used. In that case, the microscope is equipped with a scanner of 100 µm and a 100-Å radius silicone nitride tip, mounted on a cantilever of 0.6 µm. A scan rate of 2–3 Hz, a force of 10 nN, and a gain of 0.5 arbitrary units have been used. Images are then generated with Proscan. For the BioScope model, it is suggested to follow the next steps:

1. Fix the slide onto the stage by using the vacuum system provided.
2. Localize the sample with the bright field microscope using a low-magnification objective.
3. Approach the head of the microscope manually.

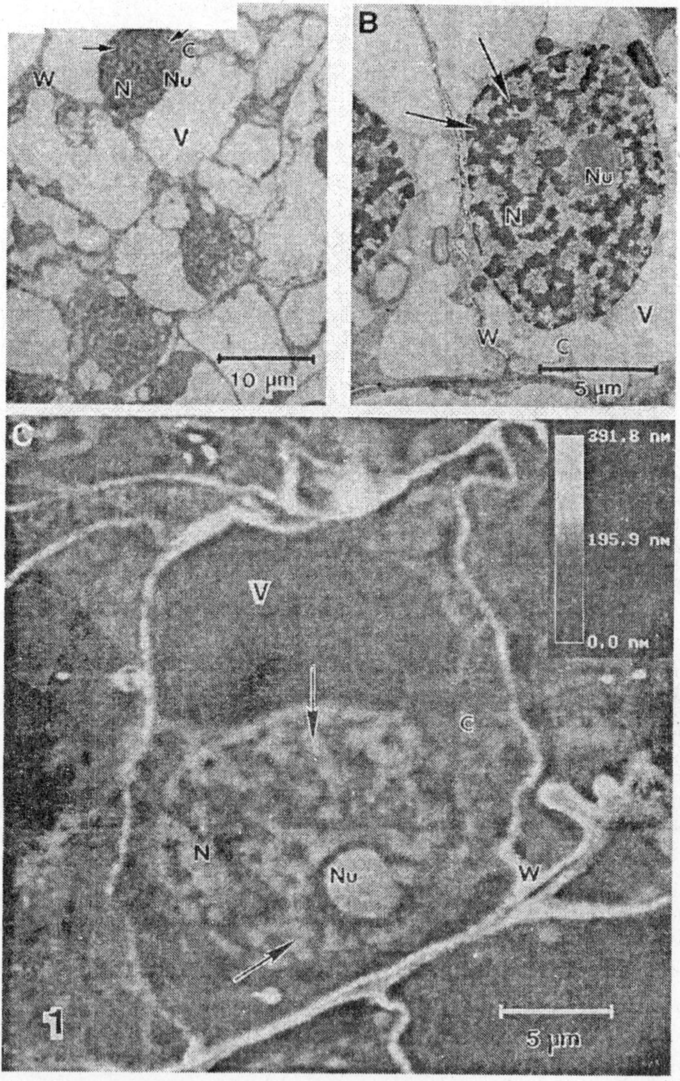

Fig. 1. Light (A), electron (B), and atomic force (C) microscopy of cell nuclei from the tegument cells from an ovary of a flower of the plant Lacandonia schismatica. *The three types of microscopes generate a similar image of the cell nuclei. However, the resolution in every case varies according to the instrument. N, nucleus; Nu, nucleolus; arrows, reticulated compact chromatin; W, cell walls; V, vacuoles; C, cytoplasm.*

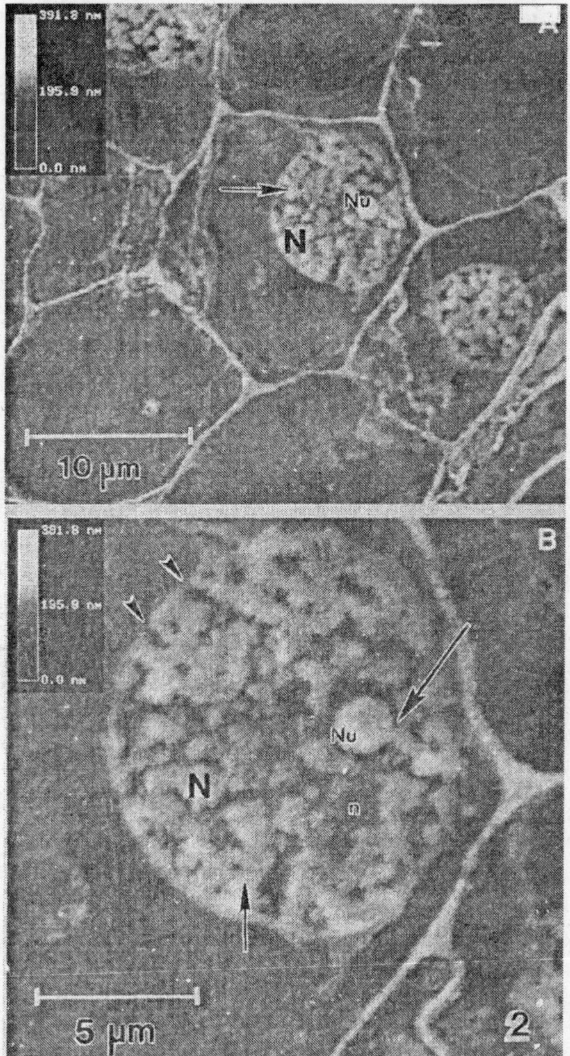

Fig. 2. Atomic force microscopy at low (A) and high (B) magnification of a L. schismatica cell nucleus. N, nucleus; Nu, nucleolus; small arrow, reticulated compact chromatin. In (B), nuclear pores (arrowheads) are observed as discontinuities of the nuclear envelope. The nucleolar organizer region (large arrow) is observed as a portion of compact chromatin strand that makes contact with the nucleolus. The nucleo-plasm (n) is seen as texturized material among chromatin strands.

4. Visualize the cantilever and the tip with the light microscope and align the tip.
5. Adjust parameters as scan size to 100 µm, scan rate to about 1.5 Hz, and data scale to about 400 nm.
6. Engage the microscope and scan the sample.
7. Generate the images using the Nanoscope IIIa software provided.

Notes

1. Regarding the epoxy resin for embedding, it is better to prepare the hardest mixture, by varying the concentration of the components such as the hardener (DDSA).
2. For obtaining sections it is better to use a diamond knife, although a glass knife also produces good quality material. Use only very clean water during the sectioning.
3. Before observation, it is recommended to place two sets of semithin sections onto the glass slide. One set is intended for staining with toluidine blue and the other will remain unstained. Under a bright-field microscope, the stained sections will be useful in recognizing the area to be scanned with the atomic force microscope on the unstained sections, which will be placed close by the stained sections. After staining, rinse the sections thoroughly with bidistilled water and air dry them.
4. Because the slide is placed sample side up, the sections face the cantilever and the tip of the atomic force microscope. However, the 10, 16, 20, and 40× objectives will image the sample through the width of the slide. Moreover, a fiberglass light will be used to illuminate the sample laterally. In an inverted microscope, it will not be possible to observe with the atomic force microscope and simultaneously perform bright field illumination according to Köhler, because the BioScope microscope is mounted on a special stage and it will interfere with the light path of a halogen lamp. Therefore, although the image visualized under the 10, 16, and 20× objectives will be of relatively good quality, when using the 40× objective, the

image will be fuzzy as a result of the width of the glass. In general, however, it will not be necessary to use the 40× objective to localize particular areas of interest.

5. A scanning field of about 100 μm is recommended at the beginning, in order to easily recognize structures at low magnification. Once a good area of interest is obtained, smaller areas can be analyzed. It is better to approach smaller areas of scanning gradually.
6. It is a common practice in our laboratory to generate images of both stained and corresponding unstained samples for comparison.
7. To operate the microscope, it is reccommended to start using the parameters as scan size accordingly (from 100 to 5 mm) The scan rate will be around 1.285, whereas the image data will remain at height and the data scale around 400 nm.

15

Imaging Surface and Submembranous Structures in Living Cells with the AFM

INTRODUCTION

In 1986, Binnig et al. revolutionized microscopy through the invention of the atomic force microscope (AFM). Subsequently, commercial instruments of this new imaging technique began to appear in the 5 yr after its discovery. In the early 1990s, Henderson et al. and Radmacher et al. both illustrated the potential of the instrument to image biological preparations in real time under nearby physiological conditions with nanometer resolution. From that point onwards the AFM has attracted biologists and the number of publications describing biological applications of AFM have grown rapidly.

A range of materials, methods, and notes have been described for imaging cultured cells with the AFM *(5)*. In general, maintenance of steady-state culture conditions involve a high degree of thermal stability (37°C ± 0.5), continual renewal of the culture medium (osmolarity of ± 320 mOsmol/kg/H_2O), and a neutral pH (7.4 ± 0.3), stabilized by the use of 20 m*M* *N*-Hydroxyethylpiperazine-*N'*-2-ethanesulfonate (HEPES) buffer and/or a flow of CO_2 through the AFM cell chamber to maintain a steady-state concentration of 5%. It is well known that fluctuations in temperature, osmotic pressure, and pH have severe affects on cell viability and structure making consistent AFM imaging difficult. Commercial liquid cells for studying biological samples under controlled conditions are available but only useful for a limited number of special application problems. Therefore, biologists designed their own setups and as a consequence a variety of homemade systems exist and differs from laboratory to laboratory when reviewing the literature.

In our AFM studies we used: 1) the Topometrix Explorer TMX equipped with a 100 µm XY/12 µm Z TrueMetrix Linearized Liquid Scanner installed on a Zeiss IM 35 inverted microscope with a homemade XY specimen stage adaptation and 2) a homemade fluid cell in combination with a heating stage. This design allows positioning of the cantilever in the optical axis of the inverted microscope, movement of the sample via the inverted microscope independently of the AFM, and minimizes cantilever drift by controlling temperature-induced variations. Time lapse images of living cells in contact or noncontact mode were obtained over a period of 2–3 h before peripheral parts of the cytoplasm started to detach from the substrate or before cell viability started to decrease as determined by the trypan blue exclusion test.

In this chapter we will give a short survey of our past research with the AFM on various (liver) cell types (Figs. 1–5). Moreover, the Methods and Notes sections address the most common problems and artifacts we encountered in AFM imaging when membranous and submembranous structures of cells wanted to visualized.

MATERIALS

Surface Coating for Nonadherent Cells

1. Collagen-S solution (Boehringer Mannheim, cat. no. 1098292).
2. Sterile water.
3. RPMI-1640 with 25 m*M* HEPES (Gibco BRL, cat. no. 52400-025).

Growth Medium and Reagents

1. For slow-growing or nondividing cells: RPMI-1640 with 25 m*M* HEPES, 2 m*M* L-glutamine (Gibco BRL, cat. no. 25030-024), 100 U/mL penicillin, 100 mg/mL streptomycin (Gibco BRL, cat. no. 15140-122), and 2% heat-inactivated fetal calf serum (Gibco BRL, cat. no. 10270-098).

2. For fast-growing cells or cells with a high metabolic rate: DMEM with 25 m*M* HEPES (Gibco BRL, cat. no. 32430-027), 2 m*M* L-glutamine, 100 U/mL penicillin, 100 mg/mL streptomycin, and 10% heat-inactivated fetal calf serum.
3. Trypan blue stain 0.4% (Gibco BRL, cat. no. 15250-061).

Apparatus

1. LifeSciences SPM Explorer TMX (ThermoMicroscopes, cat. no. SPM 4400-11).
2. TrueMetrix 100 µm XY/12 µm Z Linearized Liquid Scanner (ThermoMicroscopes, cat. no. 5180-00).
3. Zeiss IM 35 inverted microscope (Zeiss, Germany, cat. no. IM35) with a homemade XY specimen-stage adaptation.
4. 150-W Fiber light source (Schott, Mainz, Germany, cat. no. KL 150 B).
5. Linkam heating device (Linkam Scientific Instruments, cat. no. CO102, UK).
6. Video camera (Pulnix, cat. no. TM-300 CCD miniature).
7. Video monitor (Philips,, The Netherlands, cat. no. CM8833 Personal Monitor).
8. Video recorder (JVC, UK, cat. no. SR-5970E time-lapse SVHS-VCR).

Tip Specifications

1. For contact imaging: Standard silicon nitride tips (TopoMetrix SFM-Probes, Ref 1520-00) with an arm length of 200 l jm, a spring constant of 0.032 N/m and a 4 l jm on 4- l ^m pyramidal base are preferred.
2. For noncontact imaging: Standard silicon nitride tips (Topometrix SFM-Probes, Ref 1530-00) with an arm length of 100 l jm, a spring constant of 0.5 N/m, and a 4 l jm on 4- l jm pyramidal base are preferred.

METHODS

Surface Coating for Nonadherent Cells

1. Mix 30 l jL of Collagen-S solution with 270 pL of sterile water.

2. Equally distribute this solution on a 35-mm Petri dish with the aid of a melted Pasteur pipet.
3. Incubate under sterile conditions for 18 h at 4°C.
4. Rinse twice with growth medium.
5. Incubate the Petri dish for 20 min at 37°C.
6. Seed the cells and culture for at least 4 h to allow adherence and spreading of the cells on the substrate.

Contact Imaging

1. The TopoMetrix Explorer (Santa Clara, CA) equipped with a 100- l jm XY/12- fxm Z TrueMetrix Linearized Liquid Scanner and a homemade fluid cell were used.
2. The cantilever of the AFM was positioned in the optical axis of a Zeiss IM 35 inverted microscope with a home made adaptation.
3. The temperature of the instrument was set at 37 + 0.5°C by using the heating device.
4. Align the laser beam on the desired cantilever.
5. The sample was placed into the homemade sample holder.
6. Noise-free images were recorded according to the guileless of the manufacturer until the viability of the cell cultures started to decrease.
7. Typical loading forces to be used for imaging living cells should vary between 2 and 4 nN.
8. Optimize the scan rate during scanning.
9. Optimize feedback parameters during scanning

Noncontact Imaging

Fig. 1. (A) Light microscopic image obtained with a 20× objective lens mounted on a Zeiss IM 35 inverted microscope. The optical path of the eyepieces is connected to a video camera and video monitor allowing easy positioning of the cell of interest under the tip and real time observation of the AFM scanning process.

For living cells, a low resonance frequency of 38–78 kHz (depending on the appropriate spectrum frequency range of the cantilever used) and a ± 50% free-oscillation amplitude damping are preferred. Optimize feedback parameters during scanning.

Viability Assessment

1. Add 0.3 mL of 0.4% Trypan Blue Stain into the 35-mm Petri dish with the aid of a Pasteur pipette and mix thoroughly. The amount of growth medium in the 35mm Petri dish is 1.7 mL.
2. Allow to stand 5 min at 15 to 30°C.
3. Observe under the inverted microscope. Nonviable cells stain blue and viable cells exclude the stain.

IMAGING AND IMAGE INTERPRETATION DURING SCANNING

Z-Height Limitations

AFM imaging can result in artifactual images because of the limited range of the z piezo (Fig. 2). For example, in Fig. 2, the cells respond to a drug, which causes nuclear swelling in time, resulting in an increase of cell height that exceeds the limits of the z piezo (Fig. 2C). As a consequence, some nuclei have heights greater than the tip and image the structure of the whole cantilever. This can be avoided by changing the feedback parameters during imaging (*see* Note 16, Fig. 2D). Therefore, the use of the AFM is in some cases limited when the dimensions of biological structures exceed the available z displacement of the piezoelectronic scanner.

Loading Forces

Using different loading forces on the same sample will result in images with different information (Fig. 3). As outlined in Fig. 3, applying low imaging forces will result in a more accurate surface visualization (Fig. 3B), whereas increasing the force will result in obscuring of fine membranous details (Fig. 3C). However, submembranous structures become apparent when higher loading forces are applied. This submembranous imaging originates from local variations in stiffness when the AFM-tip palpates the cell membrane.

Fig. 1. *(continued)* video image the cantilever with an arm length of 200 μm is positioned above a cluster of rat skin fibroblasts. (B) Concordant AFM image of rat skin fibroblasts. Compare with 1A and notice the small cytoplasmic processes (*) as observed under the light microscope and AFM. The white bump in B correspondents with the nuclear area and the surrounding cytoplasmic area contains fine and parallel organized cytoskeleton fibers, probably representing stress fibers, 100 µm × 100 µm.

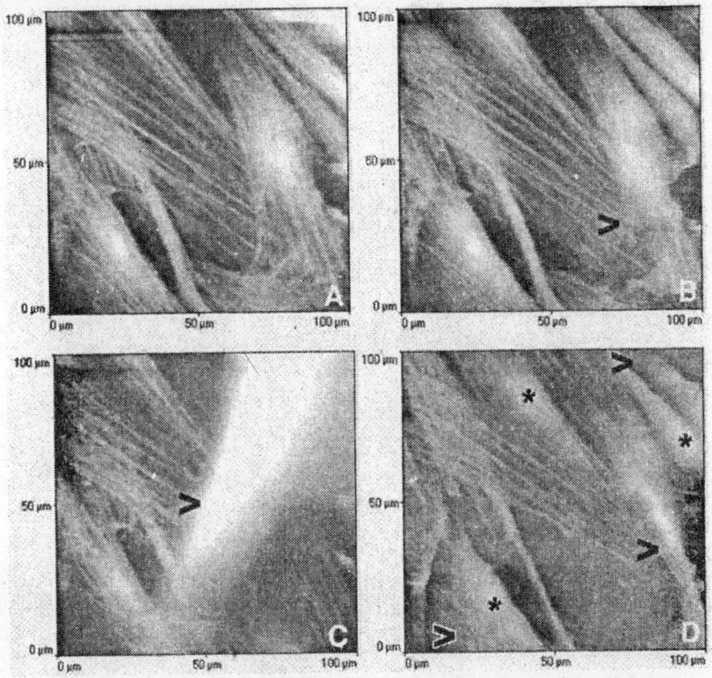

Fig. 2. Time-lapse AFM series of jasplakinolide A-induced microfilament disruption in rat skin fibroblasts. (A) Untreated fibroblasts show a parallel fiber orientation. Right after the acquisition of (A), 200 nM jasplakinolide A was added and subsequently three sequential images of the same area were recorded (recording time 1 image ~ 15 min). (B) In the middle of the image the first signs of microfilament disruption could be noticed (>). (C) The cantilever is depicted instead of the cells. Jasplakinolide A induces an accumulation of filamentous actin around the nucleus and results in an increase of the nuclear height (from 8 to 12 µm; data not shown). In this case the sample visualizes the cantilever (>), rather than vice versa. (D) Increasing the integral gain (see Note 16 during imaging resulted in an artifact-free image and reveal typical jasplakinolide-induced changes, that is, a loss of jasplakinolide A-sensitive fibers (>) and nuclear swelling (*). 100 µm × 100 µm.

A great help in interpreting results in relation to the forces used is the application of glutaraldehyde which increases the rigidity or stiffness of cells, resulting in images dominated by surface details (Fig. 3D). This can be explained by the fact that the stiffness of the cell membrane is enhanced by fixation relative to the spring constant of the AFM-cantilever, resulting in less deformation of the membrane around rigid submembranous structures.

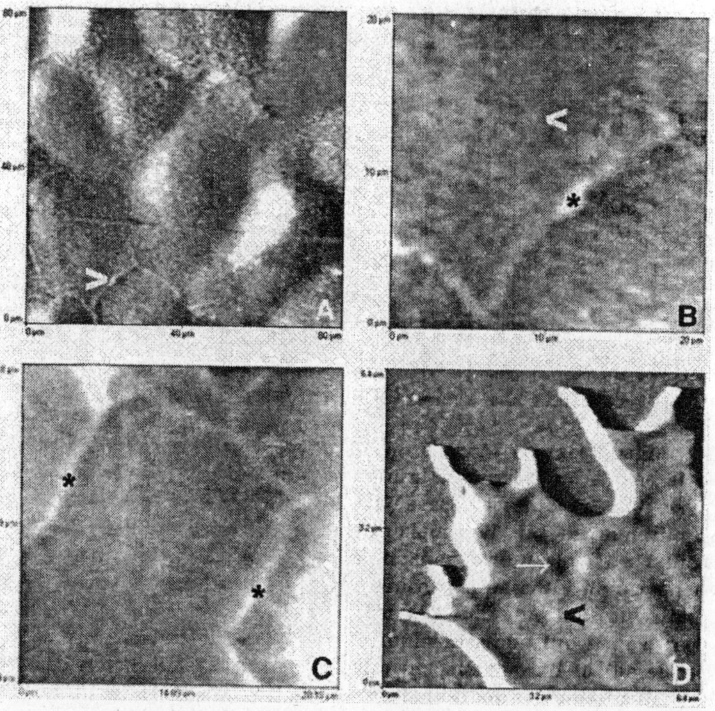

***Fig. 3. AFM micrographs of colon carcinoma cells** (see also ref. 7). (A) Low magnification showing central lying nuclei (white bumps) and clearly depicted cell borders (>), 80 lJm x 80 ljm. (B) Higher magnification of the cytoplasm obtained with a loading force of 2 nN showing fine membranous detail in the form of microvilli (>). Cell border (*), 20 ljm x 20 lJm. (C) Increasing the loading force with a factor 10 resulted in the disappearance of fine membranous detail and in clearly depicted cell borders (*), 20 ljm x 20 lJm. (D) High-magnification AFM image after glutaraldehyde fixation, confirming the presence of granular membranous elevations (>). Round membranous indentations (—>) could be visualized as well which could not be imaged in the living state, 6.4 ljm x 6.4 ljm.*

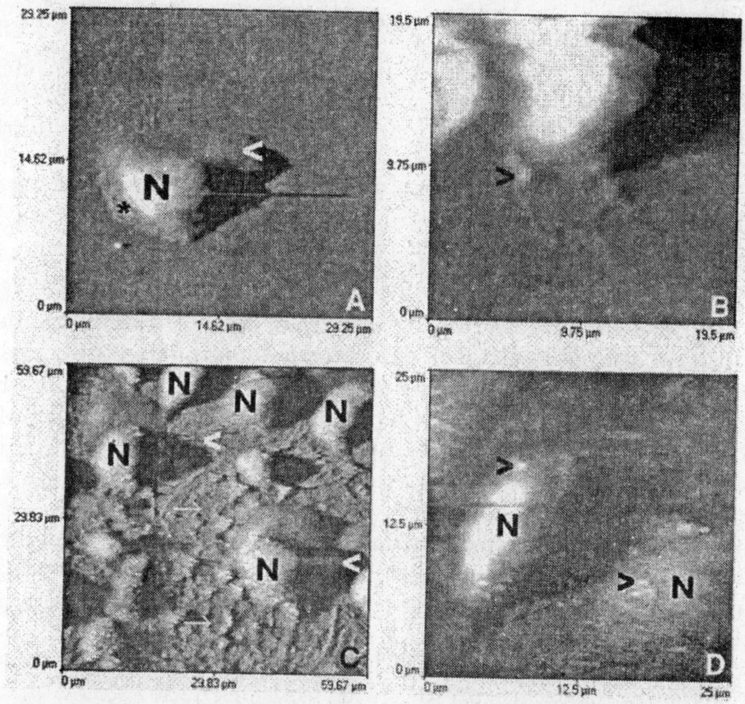

Fig. 4. AFM contact imaging of intracellular organelles in hepatic natural killer (A and B; see also *ref.* 9) *and endothelial cells (C and D;* see also *ref.* 8). *(A) Overview of one cell showing the bulging nucleus (N) and surrounding cytoplasmic margins. Note the well-attached cytoplasm (*). At the other side of the cell the scanning process apparently deformed the cytoplasm, thereby showing less detail (>). This indicates that the tip sweeps these parts of the cytoplasm along the scan direction, illustrating that these structures do not attach well to the substrate. Therefore, this part of the cell probably corresponds to the pseudopodium or leading edge of the moving cell, 29 |jm x 29 |jm. (B) Detailed image of a part of the well attached cytoplasm, showing clearly the grain-like projections (>), 19.5 |jm x 19.5 |jm. (C) Low-magnification AFM image of living hepatic endothelial cells showing well-spread cells and bulging nuclei (N) which are prominently present. Artefactual smearing by the tip (<) is evidently present. Artifactual large gaps (—>) within the cells could be noticed and probably originates from the removal of parts of the cytoplasm by the tip, 60 |jm x 60 |jm. (D) At higher magnification small white dots (>) could be observed around the nuclei (N), illustrating the presence of intra-cytoplasmic vacuoles (as known by correlative TEM studies), 25 |jm x 25 |jm.*

Tip-Induced Smearing

In general, the lateral force can wipe away or smear out surface features, whereas the constant force can deform soft biological samples (Fig. 4). Because of these tip–specimen interactions, artifactual AFM images can be obtained, that is, 1) streaks in the scan, indicating that material is being removed by the tip (Fig. 4A and C) and 2) high corrugated regions are imaged as white bumps (Fig. 4B and D), illustrating smearing or lateral deformation which is probably caused by the cantilever indenting the cell surface.

Figure 4A a nd B are AFM images of cells moving along the substrate, showing scanning-deformed membrane sheets with less detail. This indicates that the tip sweeps these parts of the cytoplasm along the scan direction, at the same time illustrating that these structures do not attach well to the substrate. Whereas, the scanning of firmly attached protrusions reveals the presence of submembranous granular projections underlying the cell membrane. In this case, the artifactual tip-induced smearing helps to interpret the activity of the cells, where the deformed membrane sheets probably represent the pseudopodium or leading edge of the moving cell. In another example, firmly attached cells show severe effects as a result of the tip– sample interactions. In this case the tip interacts with the soft cytoplasm of the cell, resulting in the formation of large artefactual gaps (Fig. 4C). However, the stiffer nuclear area facilitates imaging of perinuclear details, such as storage vacuoles (Fig. 4D).

Cell Type Limitations

It happens that the cell type of interest bears extreme phagocytotic activities (Fig. 5). In our studies we used liver macrophages, also called Kupffer cells (Fig. 5A), which have a high phagocytotic capacity for latex beads (Fig. 5B) and at the same time for the silicon nitride tip (Fig. 5C). Attempts to compose time–lapse series of images during the process of phagocytosis partly failed. Because the cells rounded up during phagocytosis and, as a consequence, the higher parts of the cells were depicted as saturated images (Fig. 5B, *see also* Subheading 3.5.1.). Moreover, during the first seconds of tip contact, as observed in the inverted light microscope, it occurs that phagocytotic cells start to react against the cantilever in an attempt to phagocytose the tip, resulting in a image of the cantilever bottom side (Fig. 5C) or in pyramidal tip images (Fig. 5D). These pyramidal tip images are probably derived from the fine cytoplas-mic protrusions, which have sharper contours than the AFM tip. In other words, the fine-edged protrusions, which are trying to embrace the tip, image the tip; rather than vice versa.

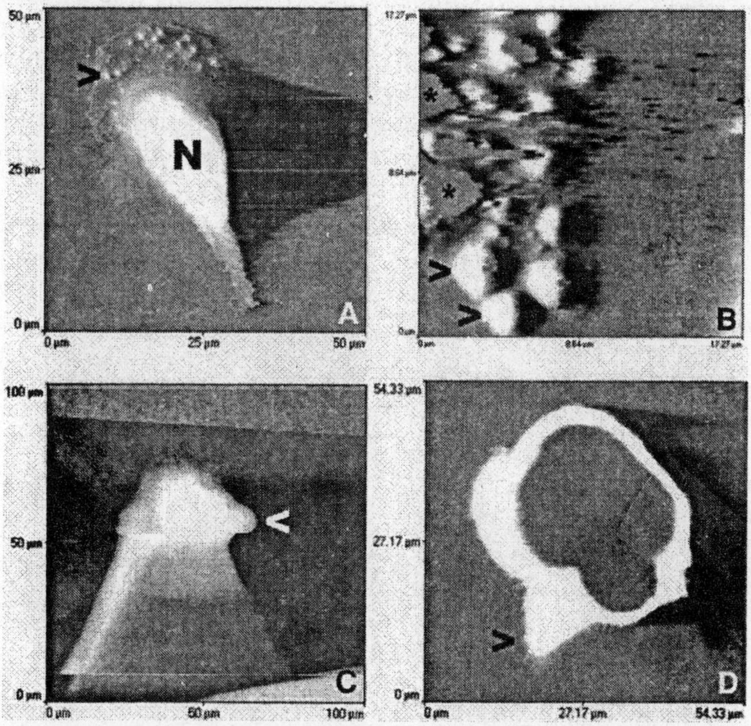

Fig. 5. Set of AFM micrographs of living liver macrophages, also called Kupffer cells (see also ref. 7). (A) Low magnification showing filopodial (below image) and lamellipodial (top image) spreading 20 min after seeding of the Kupffer cells. Notice numerous membrane projections (>), nucleus (N), 50 µm × 50 µm. (B) Living liver macrophages after phagocytosis of latex beads of 3 µm diameter. Only beads in the peripheral parts of the cells could be imaged (>). Most of the beads were depicted with saturated image information (*) because the height of the cells after phagocytosis exceeded the limits of the z piezo, 17.3 µm × 17.3 µm. (C) AFM image showing at the end of the cantilever a Kupffer cell (<), which is trying to phagocytose the tip and as a consequence loosing grip with the substrate, 100 µm × 100 µm. (D) In some cases macrophages do not attach well to the substrate and probably forms structures on top of their surface, which have sharper contours than the AFM tip. As a consequence, these structures give rise to artefactual pyramidal tip images (>), 54 µm × 54 µm.

Notes

1. Collagen solution PC-3 (ICN, cat. no. 152391) can be used as an alternative for collagen-S.
2. The pH of the medium during AFM imaging was stabilized in the physiological range of pH 7.4 by using growth medium enriched with 25 m*M* HEPES. When HEPES is used in combination with exogenous gas, it is important that the HEPES concentration must be more than double for adequate buffering, that is, 2% CO_2 approx 10 m*M* HEPES vs 5% CO_2–50 m*M* HEPES. Importantly, concentrations higher than 25 m*M* are in general toxic for cells and HEPES should be added in addition to, not in place of, sodium bicarbonate.
3. For very soft cells, such as living liver endothelial cells, which have an elastic modulus of 1 kPa, cantilevers with the lowest force constant (i.e., 10 mN/m, loading force 500 pN) should be used when the maximum resolution has to be achieved. The expected theoretical resolution for a loading force of 500 pN is 300 nm, 100 nm, and 30 nm if the sample softness is 1 kPa, 10 kPa, and 100 kPa, respectively. However, these low-force constant cantilevers have an arm length of 320 µm, making optimal laser alignment difficult for some commercial instruments and resulting in poor feedback.
4. Spontaneous adherence to and spreading on to the bottom of culture dishes is restricted to (some) hemopoietic cells, (some) tumor cell lines, and a few other selected cell types (fibroblasts) or cell lines. Therefore, the cells to be visualized by AFM have to be checked whether they are anchorage-dependent (e.g., freshly isolated rat liver endothelial cells need an addition to the substrate) vs anchorage-independent (e.g., immortomouse liver endothelial cells can be cultured directly on glass or plastic).
5. Prewarming the dishes before seeding accelerates the process of cell spreading.
6. It is advised to use well adhering and well spread cell cultures which can be easily judged by using a routine inverted microscope. Less-adherent cells can desorb off the substrate when they become in contact with the tip,

and therefore become impossible to image. However, crawling cells are an appealing topic for AFM studies, but precautions regarding the scan rate should be taken when studying them with the AFM.

7. The piezoelectric ceramics are used to generate and control scanner motion. Ideally, piezoelectric ceramic distortion is linear with applied voltage (equals linearized scanner). In principle, the present commercial scanners are corrected for nonlinear behavior. However, the scanner should be checked regularly by measuring a known sample. If nonlinearity occurs, the software of the scanner should be reinstalled or display parameters should be recalculated and adapted with the aid of a calibration sample. For more detail, all instruments contain an extended "calibration software" program in combination with a "verify calibration function." Sometimes, nonlinearities in the piezoelectric ceramics can be discovered when an image is zoomed. In this case, the image shifts from that which is desired.
8. The main advantage of imaging in liquid is the reduction of the total force that the tip exerts on the sample, since the large capillary force is isotropic in liquid. Therefore, it is important that the cantilever is completely submerged in the growth medium. For 35 Petri dishes we advise using 1.7 mL of liquid. Another obvious advantage of liquid imaging is the reduction in vibration caused by acoustic waves.
9. The combined AFM/inverted microscope is preferred when AFM imaging of living cells is to be performed. This set-up allows movement of the sample via the inverted microscope independently of the AFM and enables the user to easily locate and identify the cells or areas of interest. Moreover, correlative information is obtained and therefore improves the understanding of both microscopies, providing a limit of confidence for AFM imaging of living cells. Finally, the use of a video camera installed on the eyepieces of the inverted microscope in combination with a TV monitor and a (time-lapse) video recorder is advised, but is not obligatory. Note that laser

filters should be placed in the optical path to prevent laser light accidentally entering the user's eyes through the oculars. Moreover, to prevent light-induced damage to the cells, a broad-spectrum green interference filter should be placed in the light path of the microscope. This filter blocks the light with a wavelength below 510 nm, which is extremely toxic for cells and therefore prevents a decrease in cell viability. Because of the fact that the scan head is positioned in the light path of the microscope, a very simple way of illumination was chosen by the use of a fiber light source illuminating the close proximity of the objective.

10. It is advised to align always first the laser beam by using a blank 35-mm Petri dish (without cells) filled with growth medium before you start AFM imaging on a real biological sample. By doing so, the most common pitfalls can be discovered beforehand, such as false contact, thermal drift, oscillation, and vibration. Importantly, sometimes air bubbles might become trapped between the tip and glass sight-plate during mounting the tip on the liquid scanner. This can manifest in the impossibility to find the reflected laser spot when the liquid has been added in the liquid cell or the reflected laser spot can be found but appears to flicker, resulting in a rapidly varying sum signal from the photo-diode.
11. The Petri dish holder of the microscope can fit 35-mm dishes and is connected to the homemade XY specimen stage. The XY specimen stage and holder is made of stainless steel and could be heated by the heating device, which is placed on the saving of the XY specimen stage. To allow complete temperature equilibration it is advised to warm up the stage and holder one hour before imaging. By doing so, temperature fluctuations during imaging are avoided which can cause cantilever drift. For the same reason, the microscope, the laser and the electronic control units has to be switched on beforehand as well. Ideally, the whole instrument should be placed in a constant-temperature environment to avoid drift problems.
12. Vibration is the greatest source of image noise in AFM.

This can be easily determined when the instrument is in feedback, that is, the internal feedback signal should be stable and noise-free (equals flat line). Therefore, to obtain the highest resolution, the AFM scan head together with the inverted microscope must be maintained in a vibration-free environment. Special vibration isolation tables are commercially available and isolate the AFM instruments from the ground-floor laboratory. In addition, it is advised to place the microscopic stage on a table separate from the rest of the system. Moreover, also acoustic waves can excite vibrations in the stage and should be minimized by using a plexiglas box around the scan head for example. Sometimes, the mechanical components of the sample holder and the XY specimen stage can give rise to vibrations. Therefore, attention should be paid to make the combination of scan head and XY specimen stage as rigid as possible to avoid mechanical vibration. Vibration can also be caused by the sample holder. In this case, one or two drops of corn oil between the edges of the sample and the base plate of the microscope can solve the vibration problems.

13. To assure an optimal viability of the cells, scanning of the sample should be carried out for a maximum of 2-3 hours, after which the sample should be replaced. At the end of the experiment, the viability should be checked routinely with the aid of the trypan blue and/or the propidium iodide test. In our combined AFM-light microscope set-up, the overall viability usually drops with 5.8 + 2.1% every hour. In addition, the combined AFM / inverted microscope allows the cell and AFM tip to be seen by the optical microscope at all times during the scanning process. By doing so, the morphology of the cells during AFM imaging can be easily judged and tip-induced alterations such as detachment of removal of the peripheral parts of the cytoplasm can be easily observed. These tip-induced changes are typical morphological signs for the onset of a decreased cell viability.
14. The loading forces should be kept as low as possible under all imaging conditions. A general idea about the force

applied can be obtained by multiplying the force conversion factor (feedback) with the set point value, for example, 0.208 nN/nA x 30 nA = 6 nN. The application of high loading forces become apparent when an enhanced number of streaks are observed. This type of artifact is caused by the interaction of the tip and the sample surface and may damage your preparation. Streaking can be easily diagnosed by changing the scan direction, which should result in concomitant streaking. Streaking is more frequent when the samples are imaged under air and dry conditions resulting from the supplementary capillary forces.

15. Optimize the scan rate during scanning to avoid smearing artefacts due to tip-sample interaction. Sometimes it is possible to scan a sample so fast that the z piezo cannot react quickly enough to the motion of the cantilever. Therefore, typically the scan rate is set to two times the scan range, for example, for a scan range of 10 | jm, the scan rate would be set to 20 *[im/s*. For moving cells or less-adherent, cells the scan range should be set ideally on half of the scan range, for example, for a scan range of 10 | jm, the scan rate is 5 | jm/s. In addition, feedback parameters should be optimized as well.

16. It is of great importance to adjust feedback (proportional, integral, and derivate) and scan (rate, size, set point) parameters to optimize image acquisition. The optimum settings are largely dependent on the sample properties and therefore need to be determined experimentally. Adjusting the integral, controlling the response of the cantilever and z piezo to the largest topographical features induces the most noticeable improvement in image quality. Lowering this parameter too much, results in a smear out of the sample, whereas increasing results in a optimal image acquisition with regard to the size and shape. However, in our experience, changing the proportional gain, which controls how the z piezo will respond to fine structural details, did not affect the image quality. It is known that the effects of the proportional gain are more significant for smaller scan ranges (1–2 μm). Also the derivate gain was of secondary importance when cells

wanted to be visualized. Although when relatively large topographic cells wanted to be visualized the derivate gain acts as a stabilizing parameter. Raising this parameter reduces unwanted oscillation and allows a higher integral gain setting. The optimal value is best determined experimentally and varies from cell to cell type and scan rate used.

17. Once in feedback, the signal should be stable and have no noise. If not, move the tip away from the cells by increasing the set point (>60%). Alternatively, move the tip closer the cells by decreasing the set point (<40%). Decreasing the set point below 20% risks tip (and cell) damage.
18. Noncontact AFM takes place without physical contact between the tip and the sample and for this reason it is the AFM mode of choice for scanning soft samples (1–10 kPa), such as living cells. Another advantage of this imaging mode is the reduction of the lateral forces that can push the sample around or smear out surface features. It has to be emphasized that the mean drawback of non-contact imaging is the increased acquisition time necessary when (fast) dynamic biological processes are to be visualized. For living cells, the scan rate and integral gain are the critical parameters for noncontact imaging.

16

Atomic Force Microscopy of Protein Complexes

INTRODUCTION

Scanning probe microscopy (SPM) is a rather new family of surface studies methods, having broad applications to biomedical science. The main advantage of SPM over conventional microscopic techniques (i.e., scanning and transmission electron microscopy) is its ability to study living objects in their natural environment and, therefore, the ability to observe dynamical processes occurring to them with subnanometer resolution. Here, we will discuss one of its family members, atomic force microscopy (AFM), which has yielded fascinating results for the past decade.

AFM is a surface science technique and, thus, requires the immobilization of the studied object on a flat rigid support (substrate). A crystalline microtip attached to a cantilever is brought in contact with the investigated surface. The deflection of the cantilever is roughly proportional to the force of interaction between the tip and the surface. It can be registered by the deflection of a laser beam directed to the reflective surface of the cantilever. The electronic system maintains the constant force value, whereas the cantilever is scanning over the sample. Registration of the cantilever position normal to the sample gives the 3D surface profile. In the dynamical (or resonance) regime of registration the cantilever oscillates near the surface and the amplitude is modulated by the tip–sample interaction force. The surface profile roughly coincides with the surface of equal oscillation amplitude. (For detailed description of the AFM technique). Probe microscopy technique does not necessarily involve the sample–light interaction and, therefore, the ë/2 restriction on resolution is removed.

Although AFM can attain the atomic resolution on inorganic crystals, on biological objects it is up to date restricted to molecular one. For that reason, the main fields of AFM morphological studies of proteins is the formation of nucleic acid–protein and protein–protein complexes, oligomerization, and the organization of protein molecules in biological membranes and Langmuir films (for review).

Here, for an example of AFM observation of the protein oligomerization and complex formation, we reproduce the experiments in a system of membrane proteins (cytochrome P450–cytochrome P450 NADH-reductase), which we had previously reported and give detailed explanations of the experimental procedure and data interpretation.

Materials

1. AFM with tapping mode regime (e.g. Nanoscope, Digital Instruments).
2. Cantilevers (e.g., Nanoprobe, MicroMash).
3. Liquid cell for tapping mode.
4. Supports: highly oriented pyrolytical graphite (HOPG), mica.
5. Protein solution.
6. Tridistilled water.
7. Buffers for the protein under consideration.
8. Double-sided adhesive tape.
9. Metal discs. 10. Filter paper.

Methods

The cornerstone of successful imaging by AFM is a proper sample preparation, which can be considered as a sort of art. Preparation technique includes the correct choice of the substrate for immobilization, the adjustment of specimen concentration and buffer ionic strength, drying conditions (if necessary).

Choosing the Substrate

The main properties a good substrate should posses is flatness (a monocrystal plane would be ideal), rigidity and the studied objects' high adhesion to it. The choice of the substrate material is usually guided by the a priori information about the object's properties. Moreover, for biological research it is favorable to have a set of substrates with different hydrophobicity and surface charge of different sign. Comparing the images of the same object fixed on substrates with hydrophillic and hydrophobic surfaces one can reveal its surface properties. The strength of adhesion and the distribution on the substrate show whether the object's surface is polar, hydrophobic, or positively/negatively charged. This information is of special importance in complex formation studies, indicating which parts of molecules are in contact with each other, and which, with the solvent.

In biological SPM research, two substrates are the most widely used and are the simplest to handle: HOPG and mica. These two materials do not require any preliminary cleaning because simple cleavage (using tweezers or adhesive tape) yields atomically flat crystalline planes of considerable area (tens and hundreds of square microns) practically devoid of defects. Subsequent cleavage of the used plate allows reusing it many times.

HOPG has hydrophobic neutral surface, composed of flat terraces separated by steps. After anodization, the HOPG surface becomes hydrophilic and can be used for covalent binding of proteins.

The surface of mica is, on the contrary, polar, negatively charged, and has high adhesion to many biological specimens. Unfortunately for microscopists, a material with similar properties, but with positively charged surface, has not been found yet. To recharge mica surface, a variety of procedures has been elaborated. The simplest are based on divalent cationic treatment, others imply more complicated lipid monolayer application, or one of the most progressive techniques of silanization. Polypeptide coating produces rough surface and can be used for visualization of whole cells or organelles. Description of other possible substrates, such as silicon wafers, evaporated gold films, and Au(111) facets, can be found elsewhere.

Applying the Sample

The visualization of proteins requires special attention to be paid to the application of material onto the substrate and to related aspects of the sample purity and possible contamination artifacts. The two main techniques of application of protein molecules onto the substrate are direct adsorption and Langmuir film transformation. The latter technique, having certain advantages for amphiphylic molecules, for example membrane proteins, is described elsewhere. Covalent binding of the proteins to previously applied lipid films or silanized surfaces is specimen specific and requires a detailed description, which can be found elsewhere. Electrochemical deposition and covalent binding of protein molecules onto anodized HOPG surface are used mainly in scanning tunneling microscopy.

The direct adsorption technique implies, as can be understood from its name, a direct application of the protein solution droplet (several microliters) onto the prepared substrate. Then it is left to adsorb for several minutes, after which the unbound material is rinsed by the same buffer as containing the protein. The exact time of exposition depends on the protein concentration and the adhesion rate and should be adjusted experimentally. Typically used protein concentrations range between 1 µg/mL and 1 mg/mL. After adsorption the researcher has two choices: to place the sample into the liquid cell of the microscope, add more buffer if necessary and start *in situ* imaging or to dry the sample first and image in air.

Although the second choice does not profit all the AFM potential, it is much easier to perform, and we can recommend it to beginners or as a first step of the experiment. Dried material better adheres to the substrate and cantilever tuning is much easier as well. After drying of the buffer solution, all contaminants and salts containing the drop bulk remain on the surface, sophisticating the objects' identification. Working in liquid facilitates this problem, but on the other hand decreases adhesion stability of the molecules on the substrate. For that reason, before drying, the samples are rinsed with distilled water. The remaining water can be removed by the airflow, then contaminants of the solution will also leave the surface.

Despite all that precautions artifacts are quite often registered in AFM images of the adsorbed proteins, obtained both in liquid and air. Typically, these artifacts are all images of residual contaminants and have the shape of protrusion 3–5 nm in diameter and can be easily mistaken for protein molecules themselves. It is usually the height of the artifacts (1 nm and less) that allows one to distinguish them from the adsorbed protein molecules.

AFM Imaging

AFM measurements start with cantilever tuning. In tapping mode lateral tip–sample forces are minimized, which helps to avoid sweeping of the adsorbed material by the tip during scanning. Contact mode silicon nitride cantilevers (Nanoprobe) with a force constant of 0.3–0.6 N/m or Ultrasharp cantilevers of NSC17 series (MicroMash, Estonia) give reliable results for tapping mode imaging of protein molecules in liquid. Samples dried in airflow can be imaged with NanoSensors tapping mode etched silicon probes or cantilevers from noncontact series produced by other manufacturers. The set of parameters is individual for each system. The oscillation amplitude depends on the cantilever's resonant properties. The amplitude of free resonant oscillations A_0 can be chosen between 10 and 50 nm. The aspect ratio of loaded vs free oscillation amplitude A_S/A_0 should be kept in the interval of 0.8–0.9 for air imaging and 0.90–0.95 for imaging in liquid.

Scan rates in tapping mode are relatively low (1–3 Hz, depending on the field size). For imaging individual protein molecules, it is recommended to get several $2 \times 2\ \mu m^2$ images first (to get the idea of coverage density and choose an appropriate area) and then proceed to more detailed $0.5 \times 0.5\ \mu m^2$ ones.

Image Interpretation

In SPM the interpretation of images is a key role task, more complicated than obtaining the images. The registered objects are not "labeled" and it is normally their shape and size that serves as criterion to distinguish between them.

It is well established that apparent lateral dimensions of all AFM-imaged objects are overestimated as a result of the influence of the geometry of the probe tip, that has finite size, typically 5-10 nm, that is, the same order that a protein molecule. This can be illustrated by a simple geometrical example.

$$d = 2r^2 + 2Rr$$

Here, we roughly consider a protein molecule as a sphere

with radius *r* and the probe having a spherical tip with radius *R*. The depicted trajectory of the tip during scanning and thus the image profile has bell-like shape. Evidently, the apparent diameter *d* measured at half height does not coincide with the real diameter of the molecule. A real cross-section profile taken from the AFM image of a protein molecule (cytochrome P450) is shown in Fig. IB. Simple geometrical calculations yield the formula for this "tip-broadening" effect.

Knowing the tip radius *R* one can estimate the real diameter of molecule D

$$D = 2r = 2ylR + d^2/4- R)$$

In practice, the tip is not an ideal sphere of radius *R*, and its dimensions are usually not known precisely and differ from tip to tip. Moreover, real molecules are not spherical and can be deformed, which makes the above-mentioned calculations very approximate. For this reason, single objects' lateral dimensions are seldom used in analytical measurements, whereas vertical dimensions are preferable in calculations. A more complicated model of elliptical sample is described elsewhere *(26)*.

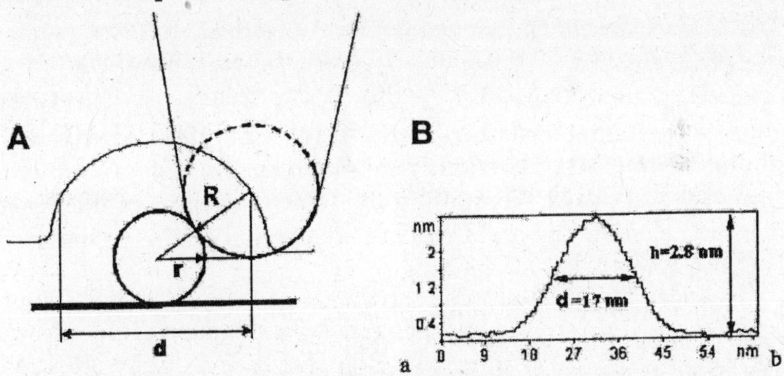

Fig. 1. (A) Illustration of the tip-broadening effect resulting from the tip geometry. R, tip radius, r, molecule radius; d, apparent molecule diameter. (B) A real cross-section profile taken from the AFM image of a protein molecule (cytochrome P450).

A successful AFM image of adsorbed globular protein molecules is a set of dots (*see* Fig. 2). Assuming the protein does not oligomerize and the shape is roughly spherical, the dots are round and of equal size. The histogram of height distribution shows a single peak (Fig. 3A). In this article we will use the molecules height, measured on the cross-section made along the scan line (*see* Fig. IB for example of cross-section). Apparent diameter measured at half height, *D*, can be used for general description or rough estimation of the lateral dimensions. If the molecules are spherical, the peak is very narrow, because all the molecules are identical. If the shape is asymmetric, the peak is widened. The molecules can adhere to the substrate with different sites, and therefore, different parts contribute to the height measurements (*see* Fig. 4).

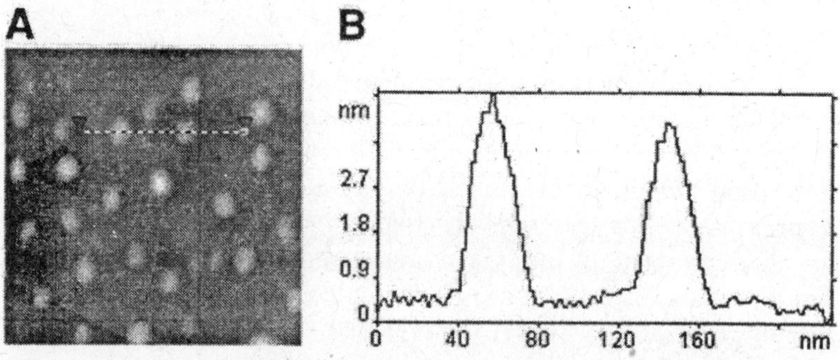

Fig. 2. Molecules of Fp in monomer form adsorbed on HOPG surface (A). Image size 0.3 × 0.3 μm²; (B) cross-section made along the marked line in (A).

If the asymmetry is considerable, for example one deals with dimers composed of two spherical globules, the histogram might show several peaks. Comparing the relative heights of the peaks one can calculate the probability of adsorption on this or that part and the free energy of binding.

When transmembranous proteins are extracted from their native membrane surrounding, their molecules form oligomers. The investigation of oligomer-ization and protein complexes formation by X-ray and diffraction methods is problematic because it requires crystallization of the sample. Meanwhile, the crystallization of transmembrane proteins is one of the most difficult crystallo-graphic tasks.

In this chapter we describe the results on AFM studies of oligomerization and complex formation for the system of two membrane proteins: cytochrome P450 2B4 and NADPH–cytochrome P450 reductase (Fp) from rabbit liver. These proteins are part of the microsomal monooxigenase system, playing the key role in the metabolism of drugs, cancerogenes, mutagens, and xenobiotics. Both proteins oligomerize when extracted from microsomal membranes.

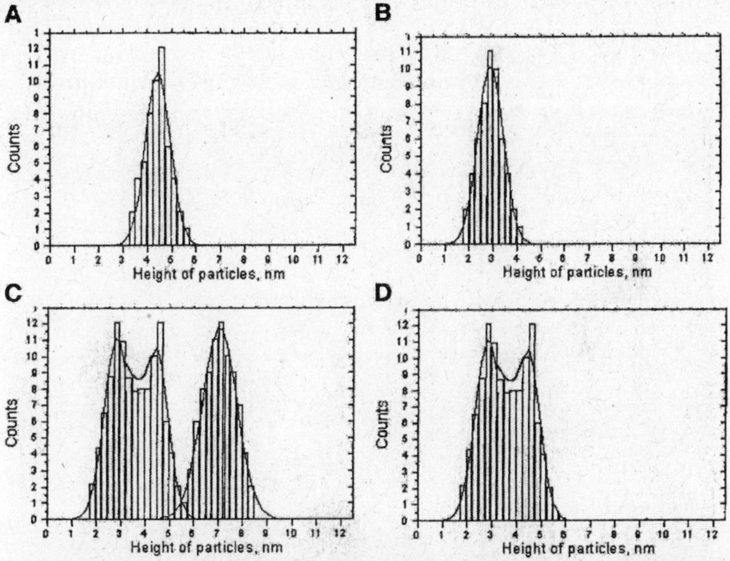

Fig. 3. Histograms of heights distribution of adsorbed proteins. (A) Fp; (B) cytochrome P450; (C) mixture of Fp and cytochrome P450; (D) mixture of noninteracting proteins.

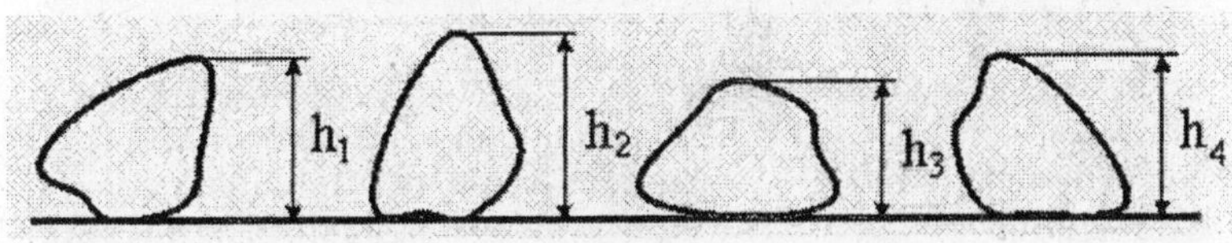

Fig. 4. Possible orientation of a protein molecule adsorbed on a flat substrae.

AFM demonstrates the efficiency of the monomerization procedure previously described *(4)* for these proteins. Fig. 2 depicting individual Fp molecules adsorbed on HOPG is a typical example of AFM image of adsorbed protein in monomer form. The distribution of heights is reflected by the histogram (Fig. 3A). The height of the molecules is 4–5 nm, the apparent diameter *d* is 20–22 nm. The histogram consists of a single peak, which indicates that one is dealing with one sort of particles. In order to confirm the fact that Fp is in the monomer form, one should compare the obtained dimensions with independent data. Despite the lack of crystallographic data, the average radius of a protein globule r in Angströms can be roughly estimated using a simple formula for spherical molecules cited by:

$$r = 0.717\ M^{1/3},$$

where M is the protein's molecular weight in Daltons.

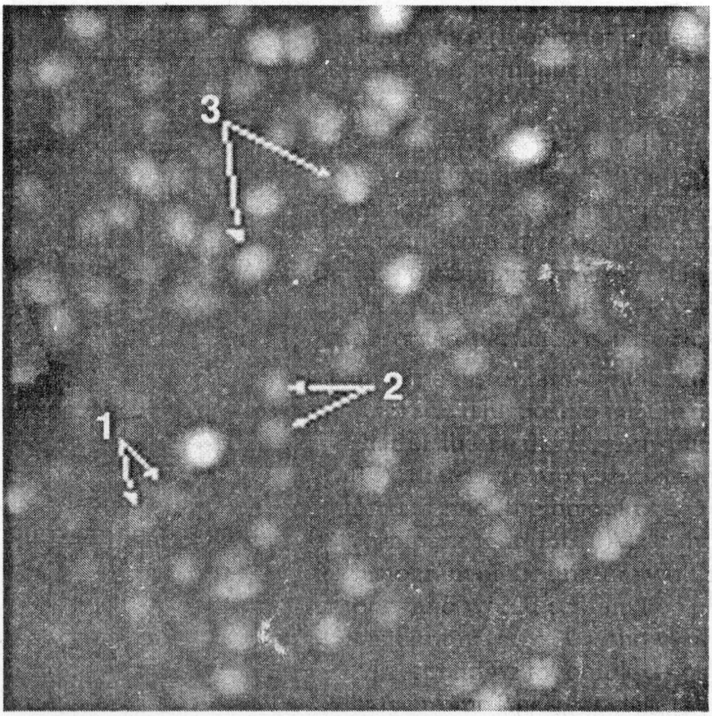

Fig. 5. Mixture of Fp and cytochrome P450 molecules in monomer form adsorbed on mica surface. Image size 540 × 540 nm^2. Arrows indicate: 1, cytochrome P450; 2, cytochrome P450 NADPH-reductase (Fp); 3, complexes of cytochrome P450 2B4 with Fp (P450/Fp).

For the Fp with molecular weight of 60 kDa, it yields $r = 2.8$ nm. Therefore, the expectation for the average height in AFM image is 5.6 nm. The registered value of 4.5 nm is quite close. The experimental value of height may be underestimated due to the deformation, caused by the tip's pressure. Thus, we can be sure that the obtained image demonstrates Fp monomers. Following the same procedure, the AFM image of cytochrome P450 2B4 on HOPG can be obtained (data not shown). The overall appearance of most globular proteins AFM images is very similar and the measurements of heights are necessary for their sizes evaluation. Height distribution histogram for cytochrome P450 is presented in Fig. 3B.

Complexes of cytochrome P450 2B4 and NADPH-cytochrome P450 reduc-tase (2B4/Fp) were obtained by mixing respective solutions of monomer proteins of the same molar concentrations. After the incubation, the mixture was diluted and applied onto the substrate.

The AFM image of the mixture applied onto mica substrate is depicted in Fig. 5. The image represents a variety of dots of different sizes, which visually seem chaotic. Because the initial mixture contained Fp monomers and cyto-chrome P450 monomers, the AFM image can represent those and, possibly, the product of their interaction, that is, complexes. The identification of the observed particles is based on the heights distribution.

The histogram of heights' distribution consists of three peaks (Fig. 3C). The position of the first peak (2.5–3.0 nm) coincides with that of the peak at the histogram for cytochrome P450 2B4 (Fig. 3B), that is, it represents cytochrome P450 2B4 monomers. The second one (the height of 4–5 nm) corresponds to Fp monomers (compare with Fig. 3A). If the two proteins did not interact, the histogram of the mixture would be the sum of the two independent ones and would contain two peaks (Fig. 3D). But the experimental histogram (Fig. 3C) shows the third peak, located at 6–8 nm, corresponding to complexes of cyto-chrome P450 2B4 and NADPH-cytochrome P450 reductase (2B4/Fp). The molecules of cytochrome P450, Fp, and 2B4/Fp are indicated by arrows with respective numbers in Fig. 5.

Using the technique described above, one can investigate the oligomeriza-tion of each of the two proteins in the absence of detergent and qualitatively reveal the dependency of the oligomerization percent versus the protein concentration and buffer ionic strength. AFM image of the oligomers of cyto-chrome P450 (Fig. 6A) is apparently similar to that of P450/Fp mixture. Individual monomers within oligomers are not resolved. For the estimation of oligomers percentage one can make height distribution histograms and determine the size of oligomers (Fig. 6B; Note 7). Determination of the number of particles in each oligomer requires a geometrical model. The choice of the model depends on the ratio of sizes and *a priori* knowledge of the molecule properties.

Notes

1. If steps are not seen in the image of 10 × 10 μm^2, either the user is very lucky to have an extremely high quality material, or the feedback system of the microscope is not working properly). When using HOPG one has to be careful about artifacts, which are now well established.
2. Because long contact with the atmosphere contaminates the substrate surface, cleavage should be performed right before the application of the sample. Before using a certain substrate for a biological experiment it is strongly recommended to get a few images of it to check for possible defects and artifacts.
3. It is strongly recommended to get several control AFM images of the buffer solution used (using the preparation technique described above) and compare them to protein molecules images, in order to reveal possible contamination arti facts. Much attention should be paid to the purity of water and chemicals.
4. It is important to bear in mind that A_0 and A_s/A_0 parameters might influence the apparent height of the biological objects imaged, introducing up to 15% error (for details, *see* ref. 24). Therefore, for analytical measurements it is recommended to use the same parameters for images one is going to compare. Using the same cantilever would be the best.
5. If the tip happens to be asymmetric, the AFM image of a spherical particle reflects the tip's shape and can be triangle, elliptical, or other. If so, the orientation of the figure is the same for all particles registered in the field. It is recommended to rotate the sample manually and see if the pattern rotates, too. If it is due to the tip's asymmetry, the orientation of figures does not change.

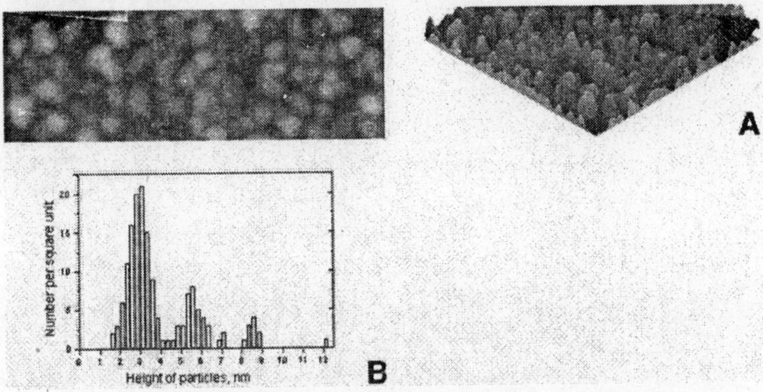

Fig. 6. (A) Molecules of cytochrome P450 in oligomer form adsorbed on mica surface, 2D and 3D images, respectively. Image size is 480 × 480 nm². (B) Histogram of height distribution shows three peaks. The first (approx 3 nm) corresponds to monomers, the second (approx 5.5 nm) and the third (8.5 nm), presumably octamers and 12–30-mers, respectively.

6. Here and further in that chapter we imaged dried samples. Such an approach is justified because the monomer–oligomer state is believed not to change upon drying. Detergents containing in the buffer often produce foam, and the bubbles do not allow imaging with a microscope with light detection of cantilever position.
7. The two proteins and their complex may have different adsorption rate to the substrate used. Therefore, one has to be careful when using the height of peaks at the histogram for the estimation of the relative part of complexes formed. If such estimations are really essential, it is recommended to calculate the amount of single molecules of one of the proteins and compare it to that adsorbed from this protein solution of the same concentration on the same area. The difference will indicate the amount, forming the complex.

17

Interfacial Monomolecular Films of Pulmonary Surfactant

INTRODUCTION

Pulmonary surfactant (PS) is a lipid protein complex secreted at the terminal airways of the lung. The material is secreted as lipid rich multilamellate bodies, which transforms into lipid–protein tubules, planar bilayers, and monomo-lecular films at the alveolar air–aqueous interface. The films reduce the surface tension of the interface and prevents lung collapse during end expiration. PS layers also act as a protective barrier against inhaled particles and bacteria and keeps the upper airways or bronchioles open during respiration. Dysfunction of PS has been implicated in various lung diseases, such as asthma, acute respiratory distress syndrome, cystic fibrosis, and pneumonia. The composition of PS is conserved in most air-breathing species; however, its high content of saturated phosphatidylcholine (PC) and phosphatidylglycerol (PG) is unique compared with other secretory materials and cell membranes, which lack these phospholipids. Specifically, PS contains significant amounts of dipalmitoylphosphatidylcholine (DPPC), palmitoyl-oleyl-PC (POPC) and PG (POPG), cholesterol, and small amounts (10%) of surfactant proteins SP-A, SP-B, SP-C, and SP-D *(1)*. It is not clear to date how this lipid–protein complex functions by forming alveolar films or barrier *in situ* because such fragile and dynamic films are difficult to preserve for traditional electron microscopy. In vitro studies have focused on model lipid–protein films of PS and also by extracting the material out of lungs and studying interfacial properties of surface tension of such material using Langmuir and other surface balances.

We have taken an approach of studying such surfactant films from lungs of normal as well as those in diseased states using a combination of fluorescence and atomic force microscopy (AFM).

Monolayer films have also become a standard model for studying lipid– protein interactions and associations in biological membranes. Models of interactions of enzymes with lipid membranes, two-dimensional crystallization of proteins, the binding kinetics of soluble proteins with a substrate, and biosensor developments have also been studied using monolayer films. Lipid films undergo a lateral phase separation from gas to fluid to gel-like phase with increasing surface packing density driven by increasing lateral pressure. The inherent changes of packing the lipid in a film undergoing lateral phase separation allows for the imaging of the structures and processes associated with formation of gas, fluid, gel, and solid domains, as well as supramolecular aggregates. The domain structures can be imaged using fluorescence and Brewster angle microscopy directly at the air–water interface. By depositing them on solid substrate using Langmuir–Blodgett technique, it is also possible to image them by AFM. The contrast in AFM image of these domains arises from differences in the molecular tilt and density of the lipids in the separate phases. Typical vertical height profiles, or topography, of films deposited on an atomically flat surface vary on the nanometer level. Thus AFM at an atomic resolution show fatty acid chains and lattice spacing of single molecules of DPPC within films. These and other AFM studies have demonstrated that monomolecular films can be used to study the molecular structure–function properties of PS and biomembrane components. This chapter focuses on the methodology for preparation and imaging monolayer films using AFM to study lung surfactant and suggests a relatively simple method to study molecular organization and disorganization (during dysfunction) of lipid–protein systems at an interface.

AFM uses a sharp tip to scan the surface of materials, which are rough at the nanometer or atomic level. Because of the interactions and deflections of the tip with the corrugated surface, real-time imaging and physical properties (friction) of such surfaces are possible in air and in liquid. However, AFM imaging of lipid films with phase transitory structures is only possible in air because the amphipathic lipids phase transition or domain formation arise from the differences in tilt of the hydrocarbon chains in air. This phase heterogeneity of packing is not observed in bilayers or monolayers, from the polar head-groups in water. However, protein–lipid interactions, binding, and crystallization processes of the proteins are better imaged in AFM in the polar head-group region because in a number of situations such processes occur in a polar environment. It is also possible to image soluble or hydrophobic proteins inserted into lipid films by AFM in air because such proteins interfere with the lipid packing. In case of dysfunction of surfactant as in respiratory disease, such as acute respiratory distress syndrome, leaked plasma proteins can enter the films from the lung aqueous interface and disrupt the surface activity of PS. We have used surfactant from a bovine source (bovine lipid extract surfactant, or BLES) and surfactant from a normal and ventilation injured rat lungs (dysfunctional surfactant), and studied them in planar films using AFM. The methods to form and study such films by AFM, and specific information about lipid packing of surfactant at an air-water interface obtained using AFM are discussed.

Materials

1. Synthetic phospholipids of high purity, such as DPPC, POPC, and a fluorescent probe 1-palmitoyl, 2-nitrobenzo-dioxo-dodecanoly phosphatidylcholine (NBD-PC) are available from Avanti Polar Lipids (Birmingham, AB). These lipids are required to measure and standardize the surface pressure-area isotherms of films and to image structure formation as a model for surfactant.
2. Commercial clinical preparations of pulmonary surfactant, such as BLES (BLES Pharmaceuticals, London, Ontario, Canada), or calf lipid surfactant extract (ONY Inc., Amherst, NY) are available. These surfactants are used mainly in clinical trials and are commercially available as a pharmaceutical product for research. They contain most of the lipid and hydrophobic protein components of natural surfactant extracted from animal lungs, except the water soluble proteins SP-A and SP-D *(1)*. We have also obtained surfactant from ventilation injured rat lungs *(18)*, however

a simple model for such surfactant can be prepared from a clinical source or similar materials can made from 10:1 wt/wt of (lipid/protein ratio) of BLES:serum protein mixtures.

3. High-purity organic solvents (99.1% high-performance liquid chromatography grade) chloroform and methanol are needed for solubilizing surfactant for film formation. Also small volumes of fluorescent probe NBD-PC in 2-5 micro liter (uL) of methanol can be directly added to the emulsion of the surfactant to form adsorbed films at the air-water interface. We have applied this method in conjunction with solvent spreading, and find both techniques yield similar film microstructures in the compressed films.
4. Doubly glass distilled and deionized water of resistivity above 18 MQ (mega-ohms) is required. The second distillation in this case can be performed using dilute KMnC4 to remove mainly organic surface-active contaminants. It is absolutely necessary for reproducible results. Surface tension of such clean water can be measured using the surface balance and should be close 72 mN/m at 23 + 1°C.
5. Cleaned glass and mica slides are required for film deposition for AFM imaging. The glass slides can be 1 cm diameter coverslips that can fit the AFM magnetic base or freshly cleaved mica of the same dimension. In case of the glass cover slips, they need to be washed first with chloroform:methanol (2:1, vol/vol) and then rinsed in chromo-sulphuric acid and doubly distilled water. Such mica and glass are to be dried in air and preserved in covered Petri dish and used directly during film deposition to avoid contaminants in the surrounding air from coming in contact with the surface on which the film is to be deposited.
6. A dedicated Langmuir–Wilhelmy surface balance with fluorescence imaging attachments was used in all experiments. Design and construction of such a balance is discussed in details elsewhere. Ours is a commercially available model (Kibron Scientific, Helsinki, Finland).

7. Scanning probe or an AFM with Silicon Nitride probes is required for film imaging. In our studies we use a DI Nanoscope IIIa (Digital Instrument, Santa Barbara, CA) scanning probe microscope with contact, tapping and tunneling mode abilities. However, other commercially available AFMs also can be used with minor alterations of the methods discussed below for films imaging. Gold-coated SiN_3 cantilevers (Wafer-113-135-22 Nanoprobe SPM tips, DI) with nominal spring constants of 0.06 or 0.38 N/m was used for contact and lateral force (friction) imaging with either a J (normal resolution) or E (high resolution) scanner.

Methods

Film Preparation

1. The Langmuir trough is filled with doubly distilled water, and the surface activity of this water is measured with a Wilhelmy dipping plate.
2. The open water interface is compressed from maximal to minimal surface area, and any surface tension drop is monitored below 72 mN/m (milli Newton/ meter) or that of a clean air-water interface. Using a suction apparatus with a sharp nozzle (Pasteur pipet), the surface contaminants are removed until the surface pressure reaches 0 mN/m or surface tension reaches 72 mN/m.
3. DPPC dissolved in chloroform:methanol (3:1 vol/vol) is applied drop-wise on this clean water interface using a micro-calibrated Hamilton syringe (10–50 µL) to form the monomolecular film. A total of 20 nM of phospholipid is applied, if the surface area of the trough is close to 120 cm^2 to give an area per molecule of DPPC to be 100 $Å^2.molecule^{-1}$. Lung surfactant or BLES films can also be formed using exactly the same technique except because BLES is a complex lipid mixture an arbitrary or average area per molecule is calculated based on the average molecular weight of the material of 750 Da. The fluid phospholipid POPC should also have similar area per molecule as those of BLES. In case of adsorbed films, similar amounts of the surfactant solution is injected from the syringe just below the air-water interface resulting in an initial surface

pressure raised to 2 mN/m. Most Langmuir surface balance software allows for automatic calculations of such film area and details of specific calculations on surfactant films are discussed elsewhere.

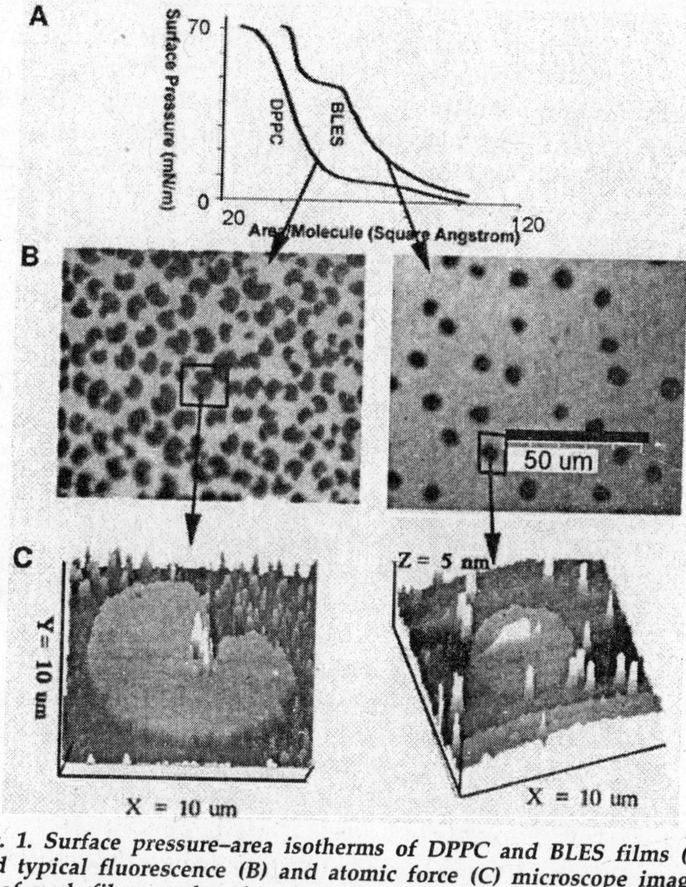

Fig. 1. Surface pressure–area isotherms of DPPC and BLES films (A) and typical fluorescence (B) and atomic force (C) microscope images of such films at the phase coexistence region of the isotherms (surface pressure of 16 mN/m). The plateau region of the DPPC isotherm in (A) suggests an expanded to condensed phase transition occurring at 4–8 mN/m. The plateau in BLES (at 45 mN/m) isotherm is possibly a higher order transition, considering that the fluid-gel transition occurs at lower pressures of 10–40 mN/m (6). The black regions in the fluorescence images (B) represent the gel or condensed phase and the lighter region the fluid or expanded phase, upon which the probe partitions. The gel regions have higher height than the surrounding fluid phase as seen in the deposited film shown in (C), and this allows for topographical imaging of films via AFM.

4. The DPPC film is rapidly compressed and the surface pressure–area profile monitored (Fig. 1). The phospholipid undergoes a two-dimensional phase transition, and this is seen in the pressure-area isotherms as a broad plateau around 5-8 mN/m (Fig. 1A). If this plateau does not occur then either the surface has not been cleaned enough, or the solvent or phospholipids contains contaminants.
5. BLES films will not show any plateau-like region below 45 mN/m (Fig. 1A), however a broad phase transition does occur between 5 and 45 mN/m. The BLES film isotherm would look similar to the one of POPC compressed at the same rate and initial area. This phase transition in BLES or DPPC films can be visually observed either by depositing the films and imaging them via AFM or using small amounts of fluorescent probe incorporated in the films and imaging them by fluorescence microscopy. Typical images taken from a DPPC and BLES films using fluorescence microscopy are shown in Fig. 1B. The black regions in the images are gel-phase domains coexisting with the homogeneously fluores cent fluid or bright regions.
6. Films are deposited on clean glass or mica slides by Blodgett deposition. The process requires a slow 0.2–10 $mm^2 \cdot s^{-1}$ up-stroke or vertically lifting of the slides from the subphase to air, termed as Blodgett deposition. The slides are originally placed into the subphase before spreading of the films from the solvent. The deposited films can be kept on a Petri dish with wet filter paper, for up to 2 days depending on the substrate of deposition. Films on glass substrates are more stable than mica, however other complications may arise using glass substrates. Comparative AFM of the deposited films (from fluorescence imaging) showing single gel domains for DPPC and BLES are shown in Fig. 1C.

Film Imaging by AFM

1. Set the mode for imaging in the Nanoscope IIIa AFM to contact mode.
2. The glass or mica slide with the films are to be attached firmly to the magnetic disk or AFM base of the E (12 μm)

or J (120 μm) scanner. Mount the disk onto the scanner head.

3. Attach the tip-cantilever to the metallic head of the AFM.
4. Adjust the laser to focus on the center of the tip. This is generally performed by observing the tip under an optical microscope or camera focused on the tip from above the scanner. General methodology for this is available in the manual supplied with the instrument.
5. Turn on the software program to online mode, and allow the tip to come in contact with the films surface. Set initial scan parameters to Integral Gain–2 and Proportional Gain–3, beginning with the 0 settings recommended for the instrument.
6. Perform a force calibration plot for the tip extracting and retracting at a z scan setting of 1 μm, set point = 0 volts, z-scan rate of 4–5 Hz, and *z* range to 1 μm. The tip deflection should read close 0.10 μm/div.
7. Open three separate channels for imaging the films in contact mode, such as channel 1, height; channel 2, deflection; and channel 3, friction. Set x/y range or scan size initially to 50 μm, and a *z* or height range to 10 nm.
8. Scan the film surface at a rate of 0.5–1 Hz (J scanner, 50–10 μm) or 2–10 Hz (E scanner 1 μm to 5 nm). The image in J scanner should appear as those shown in Fig. 2A. If no images appear at these settings and scanning method, adjust the integral and proportional gain in the software. These parameter settings and methods can be found in further details in the nanoscope operation manual and has been discussed by others in details. 9. Save the first images and exact parameters used in files, and then zoom into various areas of the films. The zooming should be performed on the center of the condensed domains, in the fluid like phase, and at the edges of the gel-fluid regions, as shown in Fig. 2B. This can reveal various microstructures observed in such films beyond the gel and fluid like phase, such as smaller microstructures or domains (possibly surfactant proteins and protein–lipid aggregates) as shown in the right panel of Fig. 2B. The gel or condensed domains

are not completely homogenous and show corrugation inside the domains as shown in Fig. 2C.

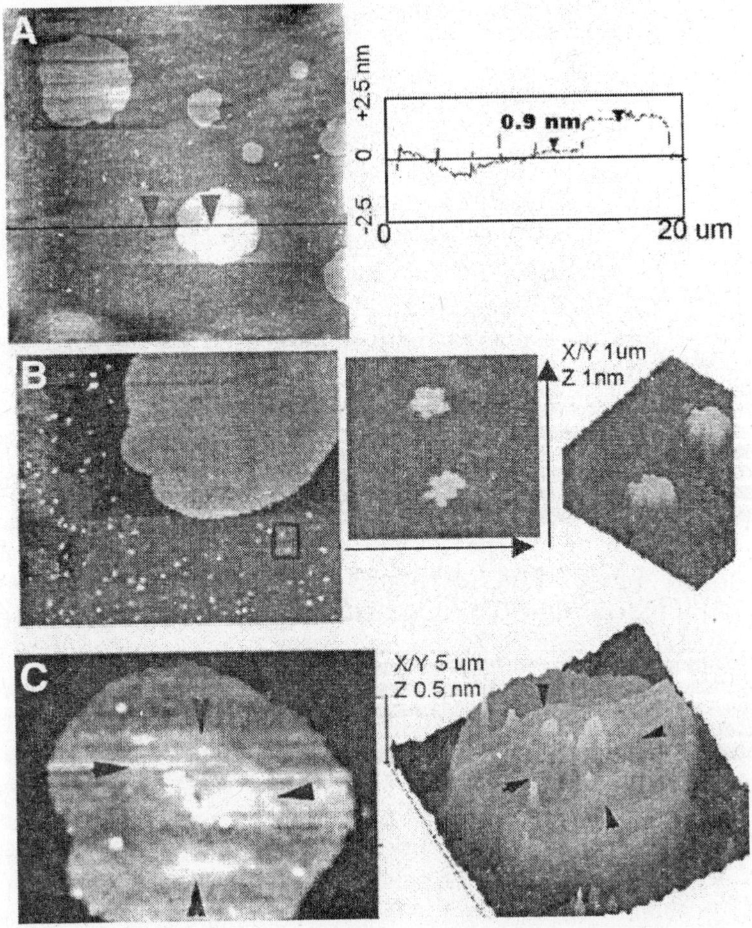

Fig. 2. AFM height mode images of solvent-spread BLES films deposited on mica. The line section analysis of the image in (A) suggests that the gel phase is 0.95 nm higher than the surrounding fluid phase, due to a different tilt of the phospholipid molecules in that phase compared with the fluid (see Note 5). The images in (B) shows that the fluid phase is not completely homogenous but has variety of micro-domains (<1 μm) as revealed by AFM and not seen in fluorescence (as in Fig. 1B). The gel domain (C) also has internal structures (region with arrows) and are possibly made with DPPC plus other saturated lipids of surfactant with differing chain lengths.

Structural Analysis and Interpretation of AFM Images

1. Change the setting of the software to off-line mode and choose the height mode image first.
2. The saved images in the height mode are then flattened. During flattening alter the z height between 10, 5, and 1 nm until the sharpest possible contrast between the gel and fluid phase is observed, as in Fig 2A the height was set to 5 nm.
3. Take a quick three-dimensional (3D) view of the image, and the gel-phase domains should be higher (brighter) than the fluid phase in such image (Fig. 2C).
4. Using the 2D flattened image perform a line section of the image in the vertical direction. The line should at least pass across a domain. The difference of this height between the gel and fluid phase should be around 0.8 to 1.3 nm. These height difference can reflect the molecular tilt of the phospholipids in the films and are typical of the gel–fluid phase height differences as in other lipid films. For BLES films the height difference is about 1.3 nm and for DPPC films it is about 0.8 nm (Figs. 1 and 3).
5. In case of adsorbed films of rat surfactant (Fig. 3A and B) other regions of protein rich areas larger than gel or fluid domains are observed (shown by arrow). These domains are possibly adsorbed soluble proteins (Fig. 3B, the circular region), which has inserted into the monolayers, and interfere with the lipid packing. The effect of lipid packing efficacy can be monitored by measuring the gel domain diameters and height from normal vs dysfunctional surfactant films.
6. Because surfactant contains a host of other lipids and small amounts of hydro-phobic proteins, various micro-heterogenous structures can be observed in the fluid-like phase. After selecting a heterogenous area, a 2D Spectral analysis using fast Fourier transform mode should be performed to observe any possible periodicity for these structures.

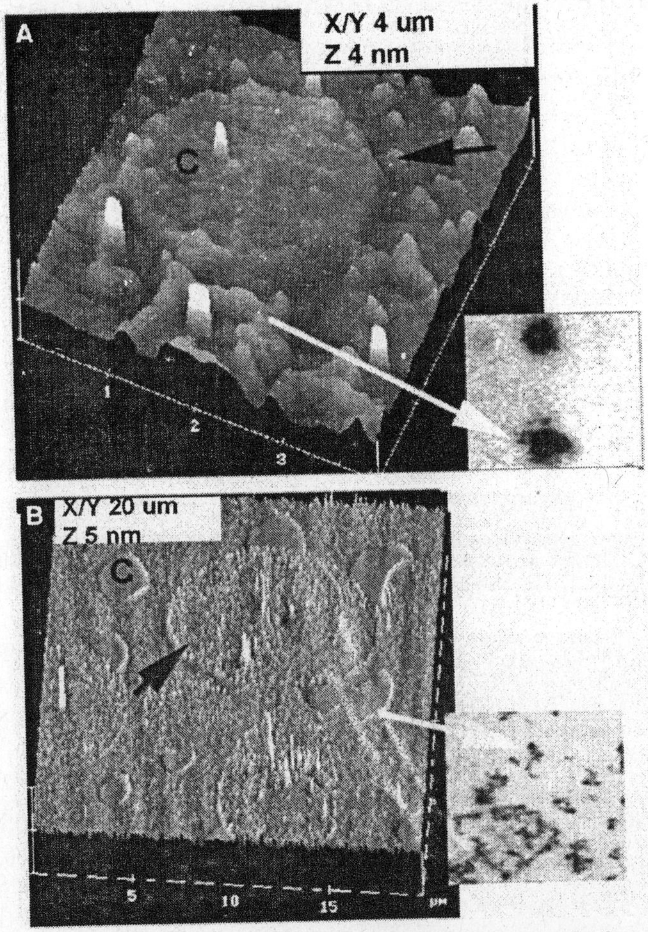

Fig. 3. (opposite page) AFM height and deflection mode images of adsorbed films of (dys)functional surfactant from rat lungs. The image in (A) is from a film of normal rat (lung–lavage) surfactant and the one in (B) from the dysfunctional surfactant, in which the rat lungs underwent hyperventilation injury. The fluorescence microscopy images shown in the right side of each panel was obtained from the deposited films. The fluorescence image suggests that the edge of the gel domains (indicated by letter C in the images) in (A) are not smooth but contain different filamentous structure and microdomain (arrow) protruding from such domains as shown in the AFM. The large circular region or domain in (B) (arrow) is of different height and deflection profile compared to the fluid or gel phase, suggesting that these regions are possibly made of protein rich phase (since the estimates of soluble serum proteins in the dysfunctional surfactant was three times higher than those from normal surfactant).

7. Save such analyzed images in a TIFF file (after the originals have been saved in the Nanoscope format) by exporting them for further display in normal image processing and display software for publication, image analysis and processing not supported by the AFM software. The AFM software normally allows for section analysis (line section, Fig. 2A), 2D Spectral and Grain size and fast Fourier transform analysis among others. Normally Adobe Photoshop, CorelDraw or NIH Imaging can allow further analysis and processing of the images.

Notes

1. For studies on interfacial film structures it is absolutely critical that a clean interface or water surface is maintained for all experiments. Although by doubly distilling (DDW) or deionizing the water using standard laboratory deionizers/ distillation apparatus, it is possible to remove the inorganic components from water (resistivity H"18 M&!). However, we have observed certain surface-active (possibly organic) components enter the water during the distillation process or from the surrounding ambient laboratory environment. From our experience a second distillation of most laboratory distilled water using dilute potassium permanganate (0.01 m*M*) dissolved in this singly distilled water before the second distillation allows for quite high quality "clean" water interface. This can be easily monitored in a sensitive surface balance, by compressing the open surface of the DDW or so called "clean" interface and monitoring any change of surface tension or pressure at maximal compression or minimal surface area. Normally if there are any surface contaminants the surface pressure would increase by 2–5 mN/m at minimal surface area. At this stage, the surface can be cleaned by suction of the interface until the surface pressure drops to 0 mN/m. This clean surface is expanded further to the original surface area of the trough to 100%, and the film materials spread on this surface. Also a DPPC isotherm of a film on such a subphase, would show a sharp fluid to gel phase transition profile (*see* Fig. 1A) in the compression isotherm between 4–8 mN/m, if the DDW

surface is clean. Surface contaminants tend to make the sharp transition more diffuse and sometimes may abolish them altogether, and can alter the typical size and shapes of the gel domains, by affecting the line tension of the gel–fluid domain boundaries.

2. It is still debated in the literature about the choice (pros and cons) of the best solid substrate to be used for film deposition for AFM. Both mica and glass surfaces are hydrophilic and therefore allow the polar regions of the films (or lipid head groups) to be deposited when they come in contact with either substrate. Although freshly cleaved mica is atomically flat and gives the best and more accurate topographical profile of film structure at an sub-molecular resolution up to the atomic level, it does produce artifacts on film structures over time due to drying, since such films are scanned in air. Glass cover slips typically have surface heterogeneity (troughs and crevices) with depths exceeding 2–5 nm and therefore allow for a certain amount of the water to be trapped in the surface with the film during deposition. From our experience the glass substrate allows for imaging of such films for up to 3 days after deposit without any major changes of the gel or fluid domain structures of the films (as observed by fluorescence microscopy in Fig. 1). However the same films deposited on mica changes structure with an appearance of microstructures (smaller domains appearing in the fluid phase) within 2—4 h of film deposit. Therefore depending on how quickly the films are scanned after deposit and the range of height resolutions required (<20 nm to >1 nm), glass or mica substrates should be chosen accordingly.
3. The parameters suggested here are the ones we use for scanning the films with our DI AFM. These can be altered depending on the instruments used; however, there are limits to how much force the tip can apply on the film before degrading the soft film structures. We find that the films can withstand forces ranging from 10 to 50 nN for imaging; however, others have used higher force >100 nN. Details of force–distance calibrations as applied to organic thin films have been discussed by others in details.

4. The imaging of the films using AFM depends only if the deposited films have the microstructures preserved as observed directly from the air–water interface using fluorescence microscopy (Fig. 1A). Other problems may occur if higher forces (<80 nN) are used during scanning, which can alter the film architecture and even remove the film from the area being scanned, by strong tip-film interactions. It is necessary therefore to image at least a DPPC film using fluorescence microscopy before and after deposit for preparing them for AFM imaging. The glass and mica slides containing the deposited film should be wiped on one surface for fluorescence imaging. Because the film deposits on both surfaces of the substrate, one of the surfaces (with film) should be cleaned by gently wiping this surface with a tissue paper or kim-wipe dipped in chloroform. This would remove the film from one surface and not allow for fluorescence from both surfaces to interfere during fluorescence imaging. From our experience we have noticed that many hours of AFM time is wasted in scanning a glass surface with broken or no films at all, since the fluorescence step was omitted. The fluorescent probe NBD-PC used below 1 mol% of the lipid do not significantly alter the film architecture, as confirmed at the air–water interface using lower-resolution Brewster angle microscopy, which do not require probe for imaging. If after using the fluorescence method, the AFM images cannot be observed, retract the tip from the surface and repeat procedure 6 through 9 on a different region of the films.
5. The condensed or gel phase in BLES or DPPC films are formed as a result of the fatty acid chain being more perpendicular to the plane of the air–water interface compared to those in the fluid or expanded phase. This allows for the gel or condensed phase to be imaged and the gel phase is considered as a tilt-condensed phase. However, AFM height differences between the gel–fluid phase do not give accurate numbers for the exact degree of tilt of the molecules in the gel compared with the fluid phase because the depression of the tip in a softer fluid

phase varies depending on the composition of this phase. In case of BLES, a number of fluid phospholipid and surfactant proteins SP-B/C are present in the fluid phase (as determined from various model film studies). Therefore the exact degree of chain tilts are actually deduced using other techniques such as grazing angle incidence X-ray diffraction from such films. These results are in close agreement with AFM results obtained by line section of the domain boundaries (Fig. 2A). The AFM images in the friction mode can indicate regions of higher tip-surface interactions or the viscoelasticity of the gel or fluid phase, as suggested in the micro-heterogenous areas in Fig. 3B, with very small height differences observed in the deflection mode.

18

Measurement of Mechanical Properties of Intact Endothelial Cells in Fresh Arteries

INTRODUCTION

Atomic force microscopy (AFM) has been successfully applied not only to the topography of biological specimens but also to the measurement of their local mechanical properties. This technique is very useful for imaging such biological specimens as cells, proteins, and DNA, because no special treatments of samples are required. The measurement of local mechanical properties in a living cell is achieved with a nanoindentation technique. The method has been applied to several kinds of cultured cells, including Madin-Darby canine kidney (MDCK) cells, myocytes, fibroblasts, and endothelial cells. However, morphology and properties are different between cells cultured in vitro on substrate and intact cells in vivo.

Vascular endothelial cells form a lining of the inner surface of blood vessel and are always exposed to blood flow. Because their morphology, internal structure, functions, and mechanical properties are closely related to each other and are strongly affected by hemodynamic factors, the measurement of the mechanical properties of the cells *in situ* is very important. Unlike the other cell types, endothelial cells are not embedded in tissues. Therefore, if a blood vessel is cut opened, we can observe intact endothelial cells and measure their mechanical properties with AFM.

This chapter explains an AFM method for the measurement of the local mechanical properties of intact endothelial cells in fresh arteries. The artery in living animals is exposed under anesthesia, and marked with stain dots on the outer surface along the axial direction.

The distances between the dots and the external diameter of the artery are measured to give the in vivo axial and circumferential lengths, respectively. Immediately after sacrifice, the artery is excised, and temporarily stored in Hanks' balanced salt solution (HBSS) at 4°C.

First, the sensitivity of the AFM system is determined using a cantilever for each measurement and a glass cover slip. Immediately before AFM observation, strip specimens are cut out from the artery. Each specimen is fixed on silicone rubber with the endothelial side up using pins at its in vivo axial and circumferential lengths.

The specimen attached to the rubber is covered with HBSS of room temperature and is mounted onto the sample stage of the AFM. After the surface topograph of the endothelium is determined, force curves are obtained at various locations in each endothelial cell. Force-indentation relations are determined from the force curves. Stiffness is calculated from the force-indentation relations.

MATERIALS

1. Living animal.
2. Pentobarbital sodium solution.
3. Gentian violet solution: dissolve gentian violet pellets in distilled water to make saturated solution; store the solution in a refrigerator at 4°C.
4. HBSS: mix 136.8 m*M* NaCl, 5.3 m*M* KCl, 0.3 m*M* Na2HPO4 · 12 H_2O, 0.4 m*M* KH2PO4, 0.5 m*M* MgC12 · $6H_2O$, 0.4 m*M* MgSO4 · $7H_2O$, 1.3 m*M* $CaCl_2$, *4.2 mM* NaHCO3, and 5.6 m*M* dextrose in distilled water, filtrate with a membrane filter having the pore size of 0.2 um, adjust the pH to 7.4, and store in a refrigerator at 4°C.
5. Silicone rubber: cut a 3-mm thick silicone rubber sheet into appropriate size (e.g., 5x5 mm). Ultrasonically wash it with acetone, ethanol, and then distilled water for 10 min for each, and dry it.

6. Stainless steel pin: cut an approximately 0.3-mm diameter stainless-steel wire into short pieces each having the length of about 4 mm and bend one end of each to an angle of about 90° to make an L-shape. Ultrasonically wash them in the same way as that for the silicone rubber, and dry.

METHODS

Resection of Arterial Segment

Exposure of Artery Under Anesthesia

1. Induce general anesthesia to an animal by the injection of pentobarbital sodium into the vein or the abdominal cavity.
2. Shave and incise the skin.
3. Carefully expose an artery, and dissect it from the surrounding tissues using forceps.

Measurement of Arterial Dimensions

1. Measure the external diameter of the artery with a caliper.
2. Dot with gentian violet on the outer surface along the axial direction at 3- to 5-mm intervals.
3. Measure the distances between the dots with a caliper.

Resection of Arterial Segment and Storage

1. Inject an excess of pentobarbital sodium solution into the vein, and wait until cardiac arrest.
2. Immediately after sacrifice, cannulate the artery with a syringe needle, and gently flush it with HBSS of room temperature to wash out blood.
3. Ligate the artery at proximal and then at distal position with threads.
4. Resect an arterial segment with a surgical scissors between the ligations.
5. Immediately immerse the resected segment in HBSS of room temperature in a Petri dish and gently wash the segment.
6. Put the segment in a bottle with fresh HBSS room temperature and store it at 4°C.

PREPARATION OF ARTERIAL WALL SPECIMEN

Cutting Out Specimen Strips

1. Transfer the arterial segment from the bottle to a Petri dish.
2. Measure the external diameter and the distance between the gentian violet dots on the outer surface. Calculate the in vivo circumferential and axial extension ratios (ratio of in vivo dimension to in vitro one).
3. Cut out rectangular specimen strips from the segment using a microscissors and a surgical blade.

Attachment of Specimen to Silicone Rubber

1. Place each specimen strip on a silicone rubber with the endothelial side up and cover the endothelium with a droplet of HBSS to keep wet.
2. Fix the specimen to the rubber with L-shaped stainless-steel pins, stretching to the in vivo axial and circumferential length.
3. Soak the specimen in HBSS of room temperature.

AFM

Mounting of Specimen on AFM Sample Stage

1. Mount a clean glass cover slip on the sample stage of AFM. Attach a cantilever to the cantilever holder of AFM and place it over the cover slip. After putting a drop of HBSS at room temperature on the cover slip and soaking the cantilever in the drop, adjust a laser beam from AFM head so as to strike the backside of the end part of the cantilever. Then, scan the cantilever or the cover slip, and obtain an image of the cover slip surface. Subsequently, determine the sensitivity of the system using the function of sensitivity measurement of AFM.
2. Input the values of the cantilever's spring constant and the above-determined sensitivity.
3. Obtain a force curve from the glass cover slip in the force curve mode of AFM, and confirm the suitability of the cantilever. Remove the cover slip from the sample stage.
4. Mount a specimen attached to the silicone rubber onto the

sample stage of the AFM, and cover it with HBSS at room temperature.

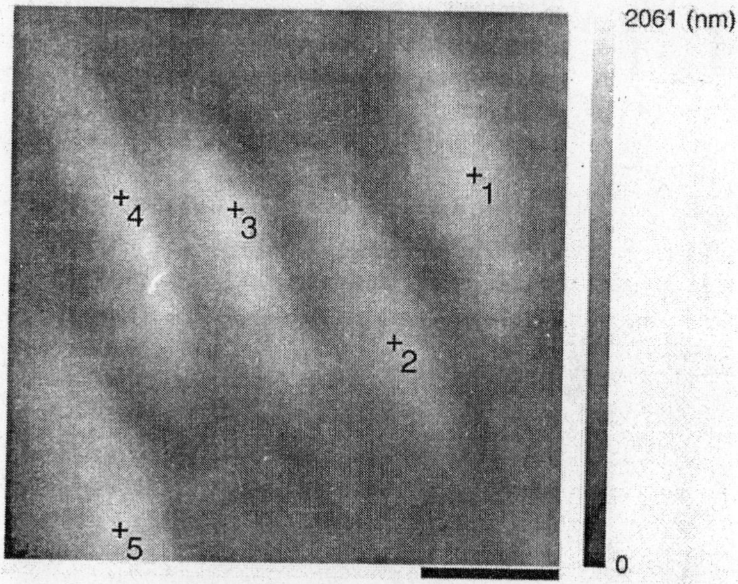

Fig. 1. AFM image of living endothelium in a rabbit abdominal aorta. Plus symbols indicate the highest points in individual endothelial cells. Black bar is 10 μm. Grey scale shows relative height.

Topography of Endothelial Surface

1. Set the *x-y* scanning range as large as possible.
2. Take a topograph of the endothelial surface in the contact mode at a low scanning rate (less than 1 Hz). Keep imaging force as low as possible to avoid the damage of endothelial cells. Change HBSS every 30 min.
3. Using the zoom function of AFM and monitoring the image, reduce the scanning size to the area of interest, and scan again to obtain a magnified image.

Measurement of Force Curve

1. Obtain force curves from endothelial cells.
2. Force–indentation relation is determined from each force curve, where indentation is obtained from the difference

between the vertical displacement of the piezo and the cantilever deflection. From the force–indentation relation, stiffness is determined.

Notes

1. The addition of very small amount of formaldehyde may help the stain attach to the adventitial surface of the artery.

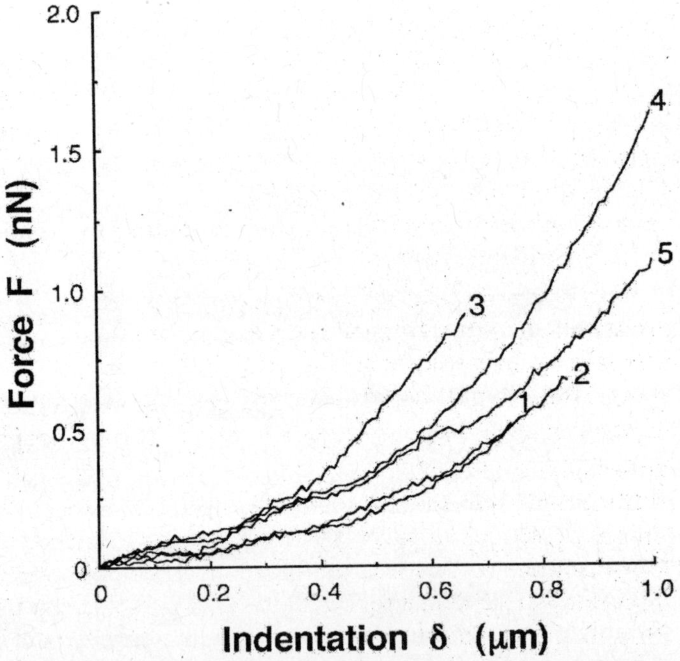

Fig. 2. Force–indentation curves obtained from the highest points in the endothelial cells shown in Fig. 1. The numbers attached to the curves correspond to the locations indicated in the AFM image.

2. HBSS is commercially available.
3. One of the tips of the wire should be made sharp so as to be easily pierced into the silicone rubber through the arterial wall.
4. Do not apply too much pentobarbital sodium to avoid respiratory failure. The dosage depends on animal species and weight. Inhalation anesthesia can also be used.

5. When exposing and resecting arterial segments, avoid bleeding as far as possible. Arteries contract when contacting with blood, which makes difficult to precisely measure the in vivo external diameter. Never grip arterial wall itself with forceps to avoid wall damage and detachment of endothelial cells. Rubbing and stretching of arterial wall also should be avoided as far as possible. Always keep the artery wet with HBSS at room temperature during the procedure.
6. If a noncontact measurement method is available, it is recommended.
7. If arteries for study are located in the legs or neck, they may be resected before sacrifice.
8. Do not flush the artery at high flow rate to avoid the detachment of endothelial cells.
9. When handling the arterial segment, hold loose fibers on the outer surface. Do not touch arterial wall itself. If blood remains inside the resected artery, gently wash it out.
10. Do not scrape the inner surface of the segment to prevent the detachment of endothelial cells. Always keep the specimen wet with HBSS.
11. Sensitivity defines a relation between the displacement of the cantilever tip (or the deflection of the cantilever) and the voltage applied to the piezo of the AFM, which is determined by pressing the tip against the cover slip using the piezo. The laser beam is reflected from the backside of the end part of the cantilever toward a segmented photodiode in the AFM head. The photodiode senses the shift of the reflected laser beam, which is induced by the displacement of the cantilever tip. The sensitivity is expressed as a relation between the output voltage from the photodiode and the voltage applied to the piezo. Thus, the displacement of the cantilever tip (or the deflection of the cantilever) can be obtained from the voltage output of the photodiode and the sensitivity. Because the sensitivity is changeable depending on the striking position of the laser beam on the cantilever, do not change the alignment of the beam until all force curve measurements are

completed. The method for the determination of sensitivity is specified for each AFM apparatus and software.

12. A spring constant is given for each cantilever. Because the actual value may be slightly different from the nominal value, it is advisable to measure or calculate it in advance. There are several methods for the determination of the spring constant of a cantilever, including a thermal vibration method.
13. A force curve shows a relation between the force applied to a specimen and the displacement of the piezo. Force is calculated by multiplying the spring constant by the deflection of the cantilever (output voltage from the photodiode). The deflection of cantilever is obtained from the sensitivity and the voltage applied to the piezo as mentioned in Note 11. The force curve of a glass cover slip is obtained from pushing the cantilever tip against the cover slip by the drive of the piezo only in z direction at a constant rate. In case the initial linear portion of the curve is not clearly observed, discard the cantilever and use a new one. The method for the determination of force curve is different in each AFM apparatus and software. A large-area piezo scanner having the maximum x–y scanning range of about 100 × 100 μm and the z range of more than 10 μm should be used, partly because the length of endothelial cells is 20–50 μm and partly because the arterial wall is not flat even if it is pinned under tension. Select a soft cantilever having a spring constant of, for example, less than 0.1 N/m and a pyramidal or a conical tip.
14. The specimen should be firmly fixed to the sample stage to obtain a good image. The silicone rubber easily adheres to the surface of the sample stage without glue.
15. A clear image is necessary to obtain a good force curve. The cantilever should be withdrawn from the specimen surface before setting the new (smaller) scanning area, because the thickness of arterial wall is not uniform and the endothelial surface is not flat. If the cantilever tip remains in contact with the specimen surface, it may scratch and destroy the endothelium, and debris from cells and/

or tissue may stick to the tip. This should be avoided because good images and force curves cannot be obtained with such a contaminated tip.

16. All the measurements should be completed within 12 h after the sacrifice of animals to avoid the deformation and structural change of endothelial cells.
17. There are various methods for the analysis of force–indentation relations.

19

Observation of Oxidative Stress on Yeast Cells

INTRODUCTION

Before the advent of the atomic force microscope (AFM), scanning electron microscopy (SEM) was used to obtain high-resolution visualizations of the surface of biological samples. Normally, to scan samples of yeast cells, each preparation was coated with a film of evaporated gold approx 20 nm in thickness. Although necessary for scanning, the application of gold to the sample resulted in distortions in its surface. In addition, the application of a conductive coating to the surface effectively masked all the information that can exist below the gold film. The AFM apparatus permits the observation of samples without the use of this mask (samples are uncoated and nonfixed).

If we compare the thickness of the gold coating to the thickness of the yeast cell wall (*Saccharomyces cerevisiae* cell wall is about 25 nm), we find that they have approximately the same dimensions, which results in loss of resolution from the surface of the cells, including any changes that might occur on the cell wall. With improvements in AFM technology, it became possible to examine the surface of many preparations at much greater resolutions than previously described. Recently, it has become possible to observe, with AFM, that the surface of the cell wall of *S. cerevisiae* contains natural undulations (rugosities) never described when SEM was used and that these cell walls contain pores along the surface that vary from strain to strain. With AFM is also possible to observe pores on membrane of others eukaryotic cells.

The ideas of pores on the surface of the yeast cell is not a novel idea, and in fact in previous studies it has been shown that it is possible to transport genetic information (plasmids or genes) to the inside of these microbes using a technique called electroporation, which involves increasing the cell wall permeability via electric pulses. Values from 2-7 kV/cm having a duration of 5 ms are used to generate pores in the cell membrane or cell wall. It is believed that pore formation generated in this manner is reversible. Unfortunately, it has been demonstrated that these values of electric pulse induce formation of reactive oxygen species and, consequently, lipoperoxidation in biological membranes (oxidative stress condition), leading to the death of a considerable number of cells. Mihai and colleagues have shown that electroporation can be used to stimulate cell growth by as much as 50% in plant cells. All such studies were very empirical before AFM technology, and until then it had not been possible to visualize if such pores were directly formed when cells were under oxidative stress conditions (induced by a chemical such as diamide or *t*-butylhydroperoxide) or stimulated by electric pulses, or conversely, if these pores were permanently in the cell wall and expand in response to electrical or chemical stimulation. Then, as with other questions, AFM solved the doubts by providing visualization of the pores and demonstrating that some strains of *S. cerevisiae* cells are resistent to oxidative stress in contrast to others.

Mechanism of Action of Diamide

Diamide, a prooxidant (Fig. 1), induces an increase in nonspecific pore formation in organelles and cells owing to the oxidation of cysteine sulfhydryl groups (SH residue) of proteins present in their membranes. The oxidation of SH to an S-S bridge also induces the formation of reactive oxygen species (ROS) and, as a consequence, lipid peroxidation in biological membranes. Therefore, diamide and other prooxidants can act as electroporation agents, inducing ROS formation and pore opening. The mannoproteins, a constituent of yeast cell wall, have cysteine in their structure and can suffer attack by diamide (and others prooxidants) and, consequently, alter the porosity of the cell wall, as observed before. This alteration of the porosity is reversible because of the antioxidant system of the cell, which is composed by adenine nucleotide in its reduced form (NADH). To induce an oxidative stress condition in the cells high quantities of prooxidant are necessary, for example, 10 m*M*.

When diamide in high concentration (10 m*M*) is added to the medium with yeast, the antioxidant system is probably exhausted, leading to an oxidative stress condition for the cells (there are no NADH molecules in the cells). As a consequence, pore closure, which is possible when NADH is present in its reduced form, is not possible.

Surprisingly, there are some yeast strains in which it is not possible to observe this phenomenon. Probably, these cells have a good antioxidant system because of higher quantities of NADH relative to the others strains.

Cell strains that produce higher NADH than others has been observed before. This good antioxidant system reduces the S-S bridge, inducing the closure of the pores. When the antioxidant system is overcome, the S-S bridge is not reduced further and, in the presence of molecular oxygen, induces the formation of ROS. These ROS can oxidize further SH residues of proteins, leading to the formation of S-S bridges and inducing a chain reaction (Fig. 2).

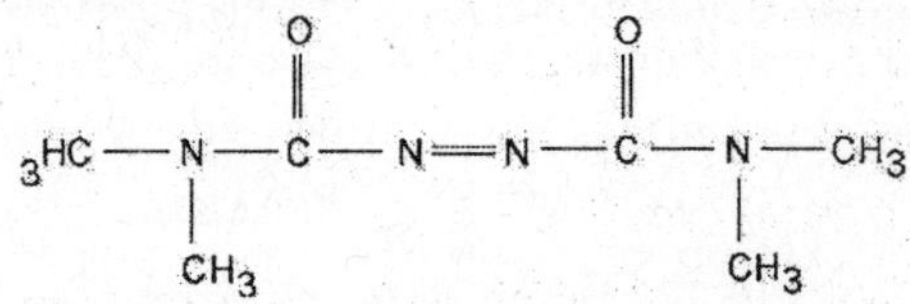

Fig. 1. Chemical structure of diamide.

These ROS induce lipid peroxidation in biological membranes, which increases the membrane permeability by opening nonspecific pores as is seen for mitochondria. By protecting sulphydryl groups from oxidation, membrane lipid peroxidation can be prevented or, at least, delayed, proving that oxidation of SH residues is directly involved with membrane permeability. If the protein has more than one SH residue in its primary structure, the oxidation of all SH present in this protein leads to the formation of an aggregation of proteins as seen before in sodium dodecyl sulfate polyacryla-mide electrophoresis; as a consequence, there is an increase in membrane permeability and pore opening. The cell walls of *S. cerevisiae* contain polysaccharide mixed with proteins. Probably, this latter controls the influx of molecules into the periplasmic space and can suffer attack by diamide or other prooxidants.

AFM as a Screening Tool

AFM technology proves to be a useful rapid screening process (45–60 min) to identify which yeast strains are oxidatively resistant, which groups of yeast are sensitive to oxidative stress, and which have pores that allow passage of macromolecules (plasmids or genes). This rapid screening tool may have direct applications in molecular biology (for example, in the transfer of genes to the interior of living cells) and biotechnology (in biotransformation reactions to produce chiral synthons in organic chemistry).

MATERIALS

1. Industrial strains of *S. cerevisiae* (lyophilized).
2. Ultrapure water.
3. 100-µL Automatic pipet.
4. Si_3N_4 AFM tip.
5. Glass cover slips.
6. Diamide (from Sigma Chemical Co.).
7. $CaCl_2$.

METHOD

Stock Suspension and Solutions

The stock suspensions of yeast cells are prepared by adding 1.0 g of dry baker's yeast to 10 mL of water (double distilled and deionized) while stirring at room temperature and the cells were metabolically active as demonstrated previously.

Control Experiment

Sixty microliters of the suspension is placed on the surface of glass cover slip and excess water allowed to dry for 12 h in air at room temperature and the cells were metabolically active as observed before.

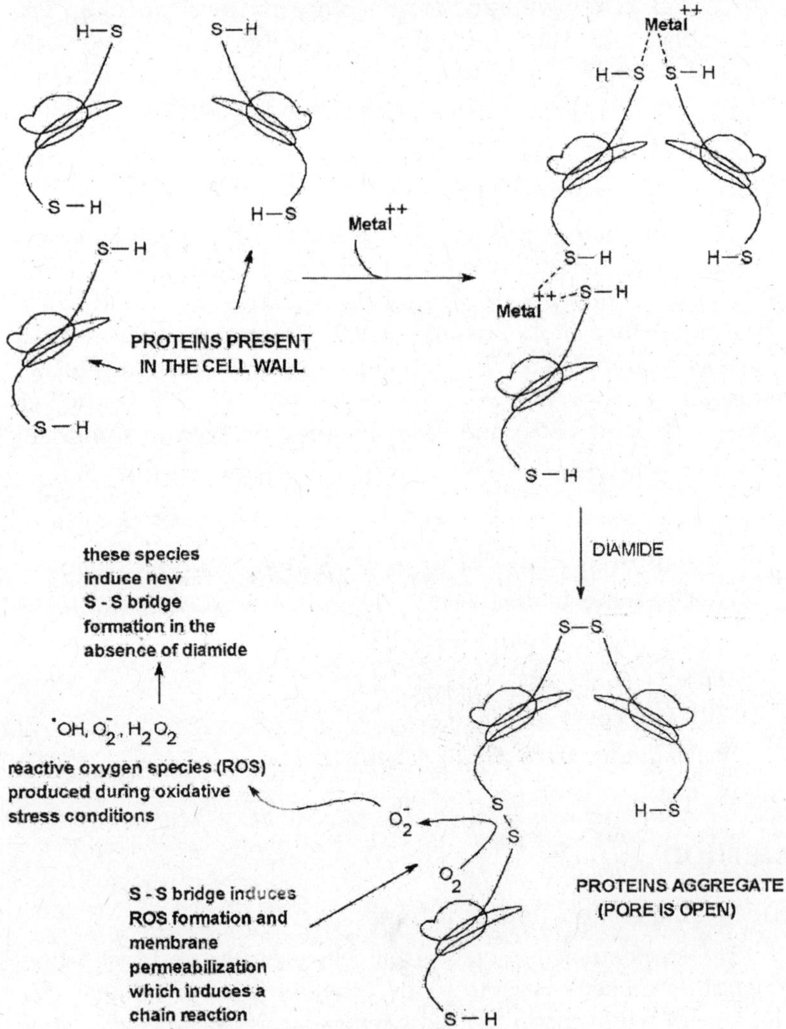

Fig. 2. Formation of S-S bridge induced by diamide and a divalent metal (metal^{++}, Ca^{2+}).

Addition of Diamide

Diamide is diluted in ultrapure water (stock solution has 10 m*M*). An aliquot of 60 μL of stock solution of diamide is mixed with 60 μL of yeast stock suspension and dried in air at room temperature from 30 min to 12 h before imaging.

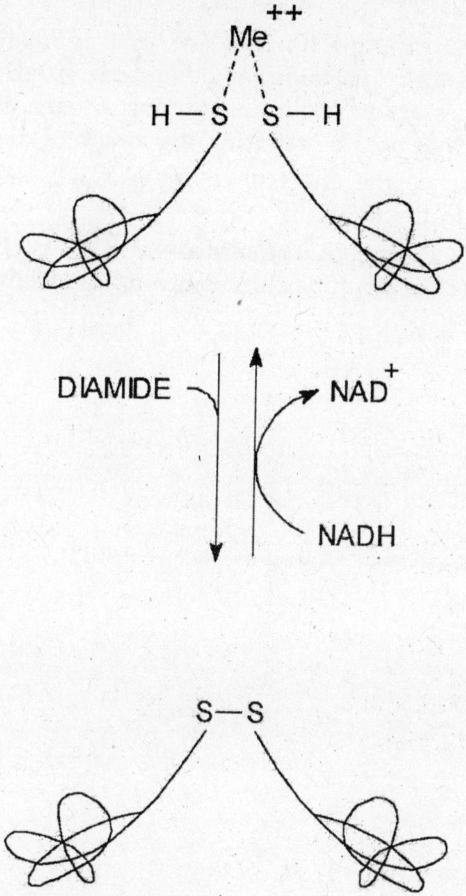

Fig. 3. Role of NADH in the reduction of S-S bridge.

AFM

A BioScope or BioProbe (AFM for biological area; VEECO Instruments) operating in contact mode is used in the experiments with Si_3N_4 Nanotips (VEECO Instruments) with a spring constant of 0.06 N/m.

In some cases, the images are low-pass filtered to remove stray scan lines. All images are collected on the AFM using a scan speed of 2.5 Hz, and all imaging is done in air at room temperature as described before *(4)*.

Notes

1. In the same sample, it seems that some populations of cells are affected by dia-mide and others are not. For this reason, it is necessary to begin by scanning an area of 50 μm^2 to know what area is suffering the attack of diamide.
2. To improve the effect of diamide, $CaCl_2$ at 300 μM is added to the medium.
3. If EDTA is present in the medium, Ca^{2+} is chelated and the effect of diamide diminishes dramatically.

20

Lymphoblastoid Cells Exposed to Low-frequency Magnetic Fields

INTRODUCTION

Since the appearance of the first pioneering article in the 1970s, particular efforts have been made to study the effect of exposure to electric and magnetic fields (EMFs) on living matter. The initial interest in radio frequency and microwaves has shifted to include the nonthermal, and essentially magnetic, effect of extremely low frequencies (ELF, 1-300 Hz), in particular the 50 and 60 Hz of the electric power system. The density of normal ELF fields is usually below 0.1 | iT, but values of 0.5 | iT may be found in front of television sets and computer monitors as well as at a distance of 50 m from 300-kV high-voltage power lines. Values up to two to three orders of magnitude higher can be found near some domestic appliances or in some industrial processes, and pulsed magnetic fields in the range 1-10 mT are used in nuclear magnetic resonance imaging and for the therapy of soft tissue or nonhealing bone fractures. The increasing interest in ELF fields has been partially motivated by epidemiological reports of an increase in some types of cancer and leukemia in children and workers exposed to high levels of ambient fields. Carcinogenesis is considered to be a multistep process consisting of initiation-promotion-progression stages. Although ELF magnetic fields are not considered sufficiently energetic to interact with DNA and initiate a cancerous process, they could, by acting on cellular receptors, affect cell proliferation and modulate the promotion or progression stage. Although exposure to a high-intensity, 50-60 Hz, magnetic field (MF) has been demonstrated to be effective in modifying the growth rate of cells, and a disturbance of melatonin rhythm has also been found-supporting the hypothesis of melatonin-mediated breast cancer development-no final evidence of a link between tumor and field MF exposure has been reported. In light of this, however, particular attention must be paid to studies performed on cells of the immune system because of its fundamental barrier action against pathogens and cancer development.

One of the main difficulties in the study of electric and magnetic field interaction with cells and in the extension of the results to different biosystems arises from the strong dependence of the response pathway on such field parameters as intensity, exposure time, wave shape, and especially the nonlinear frequency dependence. Because the plasma membrane can be regarded as a (nonideal) capacitor with a fixed frequency curve, differences in the induced effects between low frequency fields, which can modify the membrane potential, and high frequency fields, which can easily penetrate cells, are expected. Less understandable is how and why specific frequency windows (sometimes as narrow as few Hz) seem to be particularly effective in producing biophysical or biochemical cell modifications. Little is known as well about the mechanism of action when the field intensity is close to the thermal noise limit. Another complication is the known fact that present inhibitory or stimulatory behavior depends on the cell line used.

Even though no theory has completely succeeded in describing the interaction between field and cells, different models have been proposed to explain the experimental results. These models converge in attributing a crucial role to Ca^{2+} ions, suggesting a change in transport and/or in reactivity of Ca^{2+} induced by particular frequencies of the applied field. Such changes can induce a nonphysiological concentration of these ions within the cell that can trigger biochemical and genetic modifications. It is also accepted that the plasma membrane is one of the target sites of the field, so characterizing its modifications is important to understanding the nature of the interaction. A high-resolution surface-characterizing technique such as atomic force microscopy (AFM) can be a helpful tool in such a study. Recent reviews of this technique and some of its successes that justify its increasing prestige in the study of biological samples can be found in literature. A large number of experimental studies have been done both in vivo and in vitro on this subject. With regard to in vivo experiments concerning immune system cells, it has been demonstrated that nonthermal ELF MF exposure can modify the number of leukocytes, the response to inflammations, as well as the activity of natural killer cells in blood. In vitro studies of ELF MF in the range of 1–10 mT have shown membrane surface modifications as evidenced by microscopy, as well as a decrease in conductivity induced by 2.5-mT, 50-Hz exposure in K562 cells. In addition, effects on calcium fluxes, DNA synthesis, and RNA transcription have been reported, such as an increase in cytosolic Ca^{2+} concentration in HL-60 cells exposed to a 0.1-mT MF, the same effect on T lymphoblast cells *(34)*, changes in protein kinase activity in Raji cells exposed to 2-mT, 50-Hz MF *(10)*, an increase of uridine (RNA) uptake in HL-60 cells exposed to 1-mT, 60-Hz MF, and reduction in thy-midine (DNA) uptake in human peripheral blood lymphocytes exposed to 6-mT, 3-Hz (square wave) MF. It must be noted that some of the reported effects on DNA or RNA are in some way calcium-dependent or calcium-mediated so that a crucial role for calcium as target of ELF fields clearly emerges.

MATERIALS

Exposure Solenoid

The MF was generated by a sinusoidal current flowing in a solenoid (height 40 cm, diameter 20 cm) made of 600 turns of 0.2-cm (diameter) copper wires wound around an asbestos tube and placed with its axis vertical, so that the magnetic flux was perpendicular to the sample. The sample holder consisted of a water jacket (temperature regulated) container suspended at the centre of the solenoid. The current was driven by a 50-Hz power supply through a variable autotransformer generating a flux of 2 mT (root mean square [rms]) for an applied voltage of 12 V (rms). With these parameters, the surface temperature of the solenoid during the experiment is always below 30°C; no warming of the sample is possible. In order to maintain a fixed temperature during the exposure and to avoid morphological modifications induced by temperature variations, the sample was carefully maintained at a constant temperature of 37 ± 0.3°C by the temperature control system. The field intensity (measured with a calibrated Hall probe) varied within 5% of its central value inside the cylindrical exposure space.

The presence of a homogeneous electric field component parallel to the applied MF is very much debated in experiments with electromagnetic radiation because some researchers report effects induced by weak electric fields (although this still in discussion).

Surely one way to bypass such a problem is to use EMFs in which one of the components is clearly predominant with respect to the other. This is achieved in experiments with ELF fields in which the magnetic component is in the mT range. For instance, in our experiments we estimated that the residual electric component on the sample was below 80 µV/cm, which is very close to the minimum theoretical electric field to which a cell may respond, and its putative contribution can be considered negligible compared to the magnetic component.

Sample Preparation

1. Lymphocyte B cells transformed by Epstein-Barr virus.
2. RPMI (Gibco Laboratories).
3. Trypan blue dye.
4. Poly-lysine-treated glass.

The cell line used was lymphocyte B cells transformed by the Epstein-Barr virus, also known as Raji cells (from ATCC; Rockville, MD), grown at 37°C in RPMI (from Gibco Laboratories; Paisley, Scotland UK) supplemented with 10% fetal calf serum (Gibco), penicillin (110 U/mL), streptomycin (100 lig/mL), and perfused with 5% CO_2 as carbon source (which was provided also during the MF exposure). At confluence, the cells were divided into two samples and placed in 25-mL Corning flasks filled with the above mentioned medium. One sample was exposed to the sinusoidal MF continuously for 9, 15, 24, 44, and 64 h; the sham sample (control) was placed under the same conditions in a solenoid with no field. After the exposure, cell viability was determined by Trypan blue dye exclusion. This invariably showed a negligible amount of dead cells. Exposed and non-exposed cells were attached to poly-lysine treated glass cover slides, fixed in paraformaldehyde

(2% for 10 min), washed with a physiological solution, washed again with distilled water to avoid salt deposition, air dried, and finally observed in air by AFM.

This way of preparing dried samples is often used in air performed microscopy experiments to reduce as much as possible any artifact on cell morphology or viability. It allows good and reproducible results because the samples are not damaged or weakly stressed, and they remain stable for long periods of time. Moreover, to distinguish between real effects and artifacts, a control sample was prepared for each exposed sample.

AFM Instrument and Data Analysis

Our homemade AFM instrument was described in detail elsewhere; briefly, it consists of a stainless-steel unit made of two separable cylindrical supports equipped with a vibration isolating system. The lower unit contains the sample holder mounted at the top of a piezo electric scanner that allows a maximal scan size of 40 x 40 x 6 | im. Additionally an *x-y-z* motor-controlled translator is used to select suitable areas of the sample. The upper unit contains the cantilever holder, the mirror deflection system, and a four-sector position-sensitive photodiode used as deflection detector. An electronic feedback loop is used to integrate the optical signal and maintain a constant cantilever deflection during the data acquisition. Our instrument has been previously successfully used for imaging biological samples; the typical lateral resolution is about 10 nm and around 0.1 nm in the *z* direction (for a sample placed on a flat surface). In the present work, several reported images have a relatively low lateral resolution because of the large range required to observe a whole cell and because of the characteristics of our acquisition software, which allows a maximum of 256 points per row (regardless of the range of the image). However, a number of smaller images were acquired (data not shown) in which features as small as 20 nm or less were distinguishable.

The AFM measurements were performed in air (at room temperature), with the microscope working in the weak repulsive regime of contact mode with a force of less than 1 nN from zero cantilever deflection. Gold-coated Si_3N_4 (Park Scientific Instruments; Sunnyvale, CA) microlevers with a spring constant of 0.023 N/m and an expected statistical apical radius of 5–30 nm were used. Constant force and lateral friction images were acquired simultaneously with a typical scan rate of 3–4 s/row (256 points/ row).

The lateral friction-imaging mode has been found very powerful in this work because of its sensitivity to small structures, such as microvilli, that protrude from a large and corrugated surface.

The raw data were treated only using background subtraction and the repro-ducibility of data (including the absence of sample damage attributable to the measurement procedure) was successfully tested by imaging the same cells on different days. In relation to the still open question of the elastic deformation of soft biological samples because of the contact mode measurement method, it should be noted that the probe force (which was the same in all the samples analyzed) is very small and in the range commonly reported in the literature for similar contact mode experiments.

Anyway, a rough estimation of the elastic compression of a protein characterized by a Young's modulus of about 20 kg_f/cm^2 (i.e., well below the value expected in the case of a whole cell) carried out a value of elastic deformation that is largely inside the experimental error reported by us. The fit of data was calculated using the Kaleidagraph for Windows (Synergy Software version 3.08) software package.

METHODS

Image Collection and Interpretation

In Figs. 1 and 2, AFM topological and lateral friction images, respectively, of typical cells unexposed (Figs. 1A and 2A) and exposed continuously for 9, 24, and 64 h to MF (Figs. 1B–D and 2B–D, respectively) are shown. Unex-posed cells (Figs. 1A and 2A) show a characteristic domed shape and have height values (defined as top cell height minus cell background) ranging between 2.1 and 2.5 μm.

The cells appear highly structured and covered by microvilli, which are recognized both in the topographic (Fig. 1A) and lateral friction (Fig. 2A) images. Microvilli are protrusions, mainly composed of actin polymers, of the lymphocyte membrane known to be very sensitive to environmental variations and important in the recognition of inflammations sites. In the untreated sample it is also possible to see, on the border of the cells, anchoring structures named pseudopodia that are, similarly to microvilli, involved in cell migration.

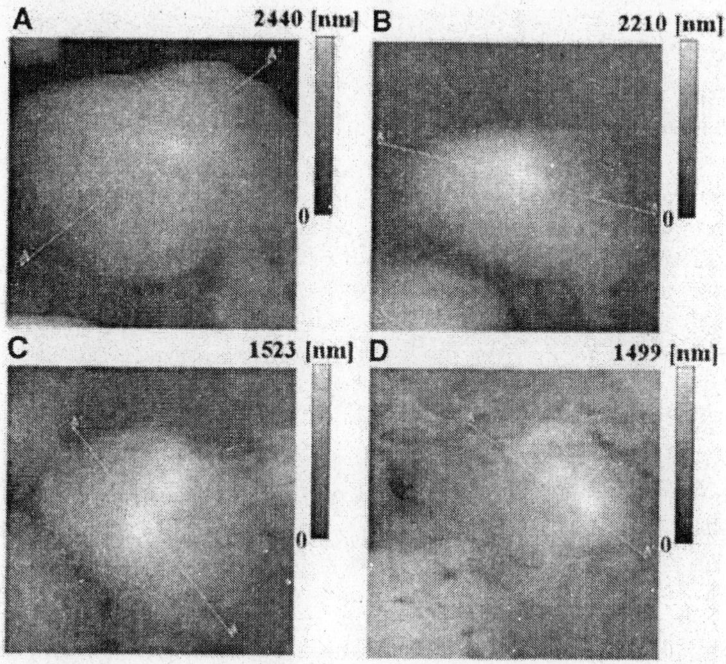

Fig. 1. Constant force AFM images of untreated (A, 10 × 10 μm) and, respectively, 9 h (B, 13 × 13 μm), 24 h (C, 14 × 14 μm), and 64 h (D, 13 × 13 μm)-exposed Raji cells. The gray scale is defined so that lighter colors correspond to higher corrugations. It is worth noting that the top height reported for the cells progressively decreases at increasing exposure. In (A), microvilli are visible as a lighter spot on the cell membrane that are no more recognizable after 9–15 h (B). At very long exposure the cell surface become characterized by several "furrows" and infolding (D). A cross-section of these cells, taken along the white line A–A′, is shown in Fig. 4.

Cells exposed for 9 h (Fig. 1B and 2B) to MF show clear differences compared to untreated cells: in this case, the microvilli have almost completely disappeared from the surface of the cells and also pseudopodia are hardly detectable.

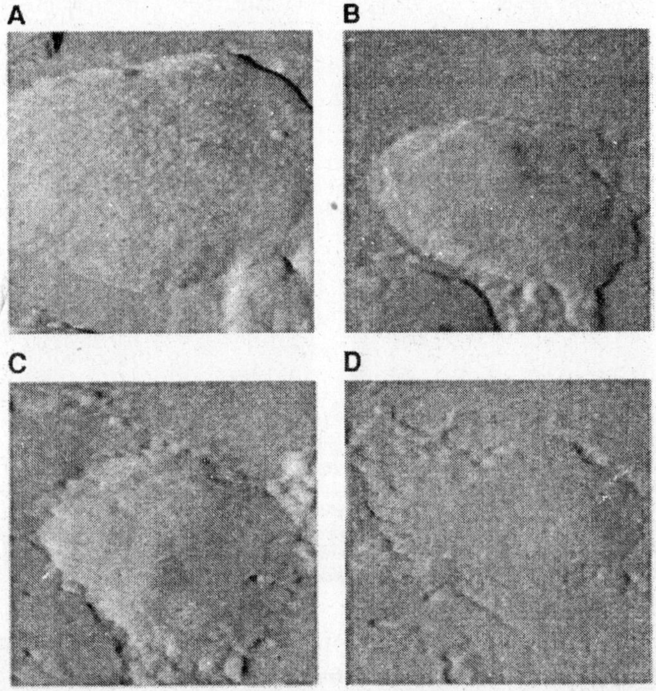

Fig. 2. Lateral friction AFM images of the same cells shown in Figs. 1A–D, respectively. Lateral friction images (collected simultaneously to topography) are very sensitive to small structures protruding from a large and corrugated surface so that they are very suitable for describing microvilli (A), as well as the fine surface modifications induced by MF exposure (for instance the pits structure in D).

Cell Membrane Features

The cell membrane, which because of the microvilli could not be directly observed in untreated samples, appears quite smooth with no structures or protrusions on the surface. In cells exposed for 15 h to MF (not shown) both microvilli and pseudopodia can no longer be recognized whereas, as in samples exposed for a shorter time, the dome shape is essentially unchanged. After 24 h of exposure to MF (Figs. 1C and 2C), a slow membrane change is still going on as revealed by the presence, in some cases, of ripples on the surface (better recognized in the lateral friction image) and by a progressive flattening of the cells. Such a membrane modification is accompanied, in cells exposed for 44 h, by the appearance of "furrows" and pit-like structures (narrow membrane infolding) that become more common in lymphoblasts exposed for 64 h to the field (Figs. 1D and 2D). These features could be considered as markers of the long exposure effect. After 64 h of exposure another important change in cell structure, namely the loss of the spherical shape of the cell, becomes evident. It is worth noting that, in some cases, this change can already be found after 44 h of exposure.

The noticeable modifications of the membrane surface because of MF exposure are shown in the high resolution (3 × 3 μm) 3D images of Fig. 3, in which the surface of an unexposed cell is compared with that of cells exposed for 9, 44, or 64 h. A comparison between Fig. 3A and 3B clarifies the effect of short time exposure, which essentially results in the loss of microvilli.

Microvilli and Adhesion

An interesting question regards the possibility that the magnetic field exposure changes the proportion of cells adhering to the substrate through the disappearance of microvilli that are involved in cell adhesion and migration: we did not find changes in adhesion although it is not possible to completely exclude such an effect.

Surface Modification Analysis

The surface modifications after longer exposure are shown in Figs. 3C and 3D, consisting in a slow "aging" of the membrane, which becomes progressively more ruffled and characterized by several narrow introflections easily recognizable in samples exposed for 44 and 64 h.

Figure 4 shows the profiles, taken along the white lines drawn in Fig. 1, of the four cells shown in Figs. 1 and 2. These data allow at least two important observations about the overall morphological changes of the cells. The first one is the progressive and relevant decrease in the maximum height of the cell with increasing exposure. During the first 9–15 h, such a decrease can be related to the observed loss of microvilli, but the residual changes must reflect modifications of other cellular structures. The second observation regards the cell's domed shape. In fact, unexposed or briefly exposed cells have high dome (Fig. 4A and B), whereas a loss of the spherical shape starts to be detectable after about 44 h (data not shown) and reaches a maximum after 64 h of exposure. Observing the profile reported in Fig. 4D, it is quite evident how this loss

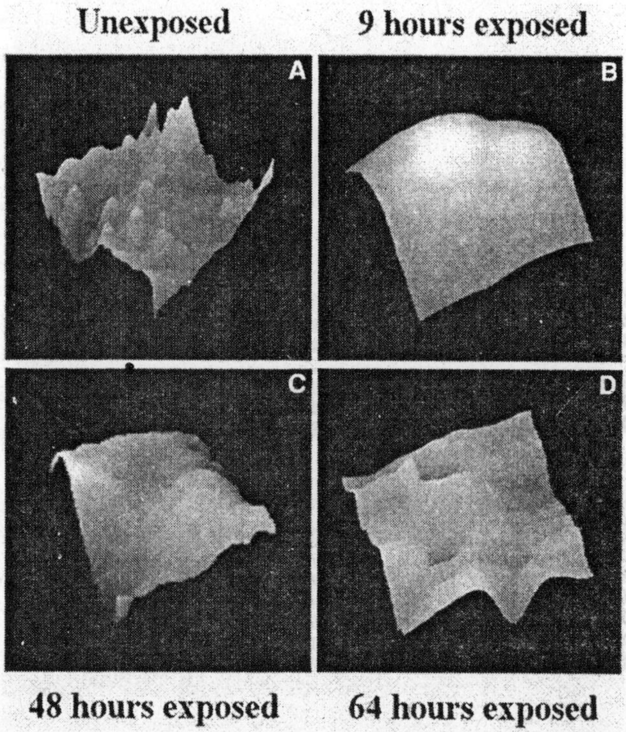

Fig. 3. Constant force images (4 × 4 μm) in a side view 3D representation of the membrane surface of an untreated cell (A), after 9 h (B), 48 h (C) and 64 h (D) of exposure to MF. The noise in these images is 0.1–0.3 nm. Microvilli are clearly visible in the untreated sample, whereas after 9 h exposure the surface shows flat and smooth. The progressive membrane ageing revealed by surface rippling and the appearance of pit-like structures is evident in images (C) and (D).

of spherical shape is the result of a weakening of the support exerted by the cytosk-eleton, the cellular structure responsible for the maintenance of the cell shape.

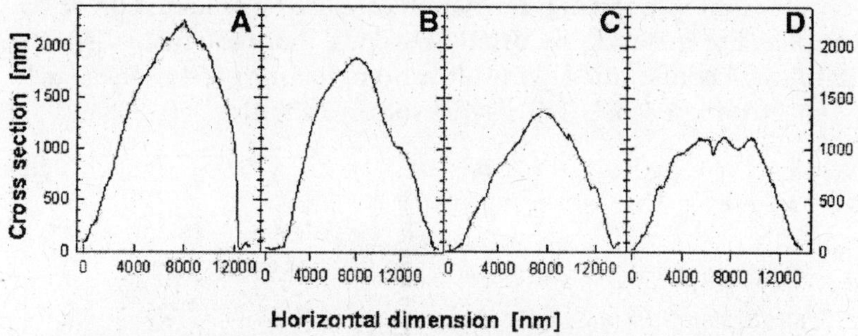

***Fig. 4.** Cross-sections of the four cells presented in Fig. 1A–D, respectively (the profiles are taken along the white line A–A'). This picture clearly shows the main morphological modifications induced by MF. They consist in the (maximum) cell height decreasing at increasing exposure as well as in the loss of cell shape taking place after long time exposure (D). The decrease in height is about complete after 24 h whereas the spherical shape is essentially conserved. At longer exposure, the residual modifications affect only the domed shape of the cell (D). A comparison of the cross-sections of (C) and (D) shows clearly that the loss of the dome shape arises from loss of support exerted by the cytoskeleton, that is, from a breakdown of this structure.*

It seems important to comment the possibility that drying might affect differently the surface structure of control and treated cells: the control cells were treated in exactly the same way as the exposed samples (except, of course, for the exposure). In this way, any difference after drying could only be caused by the exposure. A slightly higher, drying-induced, ruffling of the membrane in cells exposed for 44 or 64 h to MF cannot be excluded because of the changes in the cytoskeleton in those cells. However, we believe that this effect also, if it exists at all, has to be ascribed to the MF-induced modification of the cytoskeleton and not to the drying procedure.

Artifacts

Another point regards the possibility that while fixing and drying samples many changes might occur in the cell membrane: in the present study, we prepared air-dried samples with a weakly stressing method in order to reduce, as much as possible, any morphological artifact or effect on cell viability. Of course it would be better to study the living cells with the AFM, even though we consider that important information can also be obtained on dried samples. For instance, on dried neurones, there are many aspects that have been studied with ultra-high-vacuum techniques such as spectromicroscopy with synchrotron radiation.

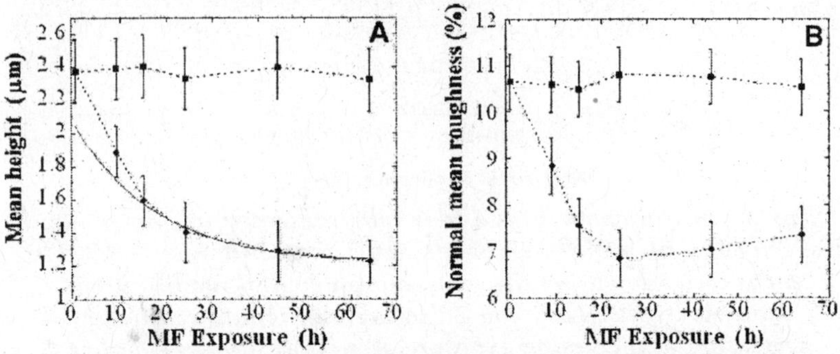

Fig. 5. Normalized roughness (B) and mean cell height (A) plotted as function of exposure time. Each point is the average of about 50 cells. In both graphs the square symbols represent the control samples and the circles refers to the exposed cells. The controls only show variations within the experimental error. In (A), we report (solid line) the fit executed on the last four data points (the ones free from effects on microvilli). The best fit was obtained with the function y = m1 + m2 x $e^{-t/18}$ with the following parameters: m1 = 1.20; m2 = 0.84; Am1 = 0.06; Am2 = 0.03. The extrapolated value of H0 (the zero exposure cells height that does not take into account the microvilli) is 2.04 um. The results of the fit are discussed in the text.

Concerning cell height and roughness, the results indicate that during the first 24 h both trends are very similar and give rise to a fast and large decrease of the parameters. We suggest that these changes are characterized by two simultaneous effects of MF on microvilli and cytoskeleton respectively. During the following 49 h, the trends become different: in fact decrease in height continues, although very weakly, while the normalized cell roughness undergoes a small increase in agreement with the progressive rippling and appearance of pit structures on the membrane surface.

Quantitative Evaluation

To allow a more quantitative evaluation of the MF-induced effects, we performed a statistical analysis of the relation between cell modification and exposure time. The results, in terms of mean cell height and normalized roughness (defined as the ratio between the height variance and the mean height value on the portion of surface analyzed), are shown in Fig. 5A and B. During the first 15-24 h, in which the main part of the MF-induced effect takes place, both graphs show similar decreasing trends. The similarity between the trends in this time frame also implies that the two phenomena of decrease in cell height and loss of structure of the membrane surface occur simultaneously. At longer exposure times, however, the trends of mean height and normalized roughness differ. The height decrease continues, although very weakly, while the roughness, after reaching a minimum value, shows a small but significant increase. This behavior is not surprising compared with the morphological data of Fig. 3, which, in fact, suggest a small increase of the roughness after long exposure in agreement with the progressive membrane rippling and formation of pit structures. This observation demonstrates the sensitivity of our statistical analysis to fine morphological modifications.

In the graph of cell height two different rates of variation are recognizable: a faster one during the first 15 h and a slower one after longer exposure. Because the height decrease continues even after the disappearance of microvilli, an important but time-limited phenomenon, it is clear that the MF acts also on other cellular structure. This structure is the cytoskeleton, subject to a slow but continuous modification. During the first 15 h, the superposition of these two effects causes the faster rate of height variation that is one of the most important results reported.

To avoid the possible interpretation that the data results from a decrease in cell volume, we measured the (apparent) cellular volume individually using an approximation of the cells as spheres or hyperboloids. The results (not shown) reveal volume changes within the experimental error, which means that as height decreases the cells become progressively wider at increasing exposure.

Role of Calcium

An interpretation of the effects we observed brings into play the role of calcium. Microvilli and pseudopodia are in fact dynamic structures, mainly composed of poly-actin filaments, that can be rapidly created and destroyed because of Ca^{2+} concentration fluctuations that are known to be induced by exposure to MF. Because actin is present both in the microvilli and in the rest of the cytoskeleton, it is reasonable to believe that the cytoskel-eton undergoes the same depolymerization effect observed in the microvilli. However, in the case of the cytoskeleton, the effect is expected to be smaller because of the rigidity of this structure, and it can be unequivocally identified only after long MF exposure (i.e., when the microvilli have already disappeared).

Estimation of the Effect Induced by MF on the Cytoskeleton

A possible, although rough, estimate of the effect induced by MF on the cytoskeleton during the first 15 h may be attempted by fitting the last four points of the height curve, which are the ones completely free from effects on microvilli. The result of the proposed fit, performed with a simple mono-exponential function (Fig. 5A and its caption) show the two rates of the phenomenon. In fact, the best fit obtained, which does not take into account the microvilli, describes very well the range of 13–64 h (for construction) but clearly indicates a rate of height variation slower than the experimental one in the first hours of exposure. The difference between these two rates of height variation can be ascribed (in large part, at least) to the effect on microvilli.

The use of the fit also allows extrapolation of a zero exposure height value (H0) that takes into account only the effect on the cytoskeleton. The total height variation during the first 9–15 h of exposure can be written as follows:

$$"H_{tot} = "H_c + "H_m \quad (1)$$

where $"H_{tot}$ is the total height variation; $"H_c$ is the contribution to the height variation because of the cytoskeleton and $"H_m$ is the contribution to the height variation attributable to the microvilli.

At $t = 0$ $"H_c$ can be estimated by the zero exposure fit extrapolation $H0$ (equal to 2.04 μm) and the contribution because of the microvilli ($"H_m = "H_{total} - "H_c$) results to be about 0.34 μm (with an error of 0.09), a value close to the 0.4 μm suggested by Knutton et al. *(47)*. It is worth noting that this value should be considered as an independent estimation of the size of microvilli in the sample analyzed.

Comparison With SEM and Fluorescence Microscopy

It is worth noting that our results are in agreement with previous data of Santoro et al. *(10)* that report, by scanning electron microscopy of lymphoblastoid cells, the loss of microvilli after 72 h exposure to 50 Hz, 2 mT MF and also show, by fluorescence microscopy analysis, a rearrangement of actin subsequent to (72 h) exposure. This supports the interpretation of our data with regard to effects on the cytoskeleton. In the same paper the authors also provide Laurdan spectroscopy evidence of a membrane fluidity variation that can be related to the progressive modification of the membrane leading to the appearance of rippling and pit-like structures reported here.

In this view, our data enable us to extend, and roughly quantify, the detection of cytoskeletal modifications during the first hours of exposure and to add a 3D description of the MF-induced changes. We can also introduce an experimental correlation between exposure time and morphological parameters such as cell height, shape, membrane roughness, and carry out the variation kinetics of these parameters to determine markers of long MF exposure.

COMMENTS

The AFM images reported here demonstrate the existence of an exposure-dependent MF-induced morphological effect on immune system cells (Raji). This effect can roughly be divided as follows: within the first 10–15 h there is a large decrease of cell height and roughness related to the disappearance of microvilli with a minor simultaneous effect on the cytoskeleton. At longer exposure time the plasma membrane appears to become completely free of microvilli and the weak residual variation, which can be completely ascribed to the cytoskeleton, leads to a less domed and wider cell shape and to the appearance of ripples and pit structures on the membrane surface.

The reported data allow us to speculate that in such treated cells some functional alteration occurs (for instance in cell motility or target recognition). However, the very large diffusion in intensity and frequency of the MF used in our study requires caution in drawing conclusions about a possible health hazard.

Further information about actual cell damage induced by MF will come from the characterization of the degree and the kinetics of reversibility of the morphological changes, and also from the study of the correlation of morphological changes to biochemical modifications and cellular dysfunction. It could also be interesting to establish the threshold value of the field intensity below which no morphological modification is detectable. Specific studies are in progress to extend the experiment to different cell lines that, because of the known specificity of the MF-induced effect, could present a different response pathway.

21

How to Build Up Biosensors with the Cantilever of the Atomic Force Microscope

INTRODUCTION

With the advent of the atomic force microscopy (AFM), the study of biological samples has become more realistic because, in most cases, samples are not covered or fixed and this makes it possible to observe them while alive. This advantage of the AFM prompted a new invention: nanobiosensors using the cantilever (probe) of the AFM, which made possible the observation of specific molecules (including medications) as they enter or exit living cells.

The nanobiosensor is the smallest biosensor in the world and measures about 100- l im long (about the width of a hair). Beyond sensing the area of interest, this biosensor also makes possible a real-time image of exactly what is occuring on the cell.

Use of the AFM as a Heat Detector

An apparatus for measuring variation of temperature is the thermooptical detector. A reaction cell, whose bottom is a thin gold film (with immobilized enzyme), is immersed into a CCl_4 phase where a probe beam is passed. The heat (from reaction between the enzyme and its substrate) of the water phase is transferred to the gold film and the CCl_4 phase, and thus a temperature gradient is generated in the CCl_4 phase. This temperature gradient induces deflection of the probe beam and the result is registered as a graph.

Because AFM has almost the same parts as the thermooptical detector (laser beam, lens, and photodiode, if an enzyme was immobilized on the cantilever, one can hope that the temperature gradient generated by reaction heat (between the enzyme and its substrate) could induce a deflection of the cantilever, transforming the AFM into an apparatus that could reveal the presence of specific molecules.

As mentioned before, an AFM equipped with an enzyme-coated cantilever (nanobiosensor) can visualize living cells, so it should be possible to detect specific molecules being absorbed by living cells and to see this happening. These assumptions were shown to be valid by experiments made with the first nanobiosensor, which had been made to detect the absorption of glucose molecules by *Saccharomyces cerevisiae* cells.

The deflection of the cantilever is explained by the physical phenomenon occurring when two different metal blades (each with a different coefficient of dilation) are bonded together. This is called a bimetallic system, and the presence of heat makes the blades bend. Another illustration of the phenomenon is the working of the taillights of an automobile: a bimetallic blade is heated until it bends and produces a braking light.

The cantilever of AFM is not a bimetallic blade but can work as one because it consists of two different materials (a metal and semimetal) with different dilation coefficients: gold and silicon. If a biochemical (or chemical) reaction takes place near this so-called bimetallic blade, it is possible to produce a deflection in the cantilever. This latter produces an enormous deviation in the direction of the laser beam, which is recorded in real time by the computer.

Beyond observing the absorption of glucose molecules by living cells, it was possible to visualize the inverted phenomenon: the release of specific molecules from living cells immobilized on the cantilever surface, in the first instance of this technique, the enzyme alcohol dehydrogenase type II, which transforms ethyl alcohol into glutaraldehyde and releases heat in the process.

Other biosensors were made using the enzymes superoxide dismutase and catalase, proving that this methodology is valid for all enzymes or molecules that release heat from their reactions.

Identifying Defective Genes

Tiny variations in the genetic code are what make us unique. Changes in one chemical unit in a gene's sequence, called single nucleotide polymorphisms, can influence an individual's risk for disease. Comparing variations will help identify defective genes and may lead to better diagnosis and treatments.

If a cantilever is coated with DNA strands, it is possible to detect genetic variations or single nucleotide polymorphisms when an individual's DNA is tested against them.

In a recent report, scientists from IBM demonstrated that when DNA strands are immobilized on a cantilever surface, it is possible to identify the complementary strand of the DNA in solution. When the immobilized and complementary strands find a match, the heat evolved from the process makes the cantilever bend. This nanomechanical property transforms the cantilever into a DNA chip, making it very useful in identifying mutations.

Clinical Analysis

The immobilization of para-nitrophenylphosphate on the cantilever can be used to identify cancer in the bones, prostate gland, and ovary. Para-nitro-phenylphosphate is the substrate of the acid phosphatase enzyme and it is currently used in clinical analysis. The enzyme is released by tumor cells into the extracellular medium.

MATERIALS

Enzymes (such as glucose oxidase, catalase, superoxide dismutase, or alcohol dehydrogenase type II) are obtained from Sigma Chemical Co. All reagents are of analytical grade. Water is double distilled and deionized.

The industrial strain of *S. cerevisiae* is commercial. The stock suspensions of living cells are prepared by adding 1.0 g of dry baker's yeast to 10 mL of water (double distilled and deionized) while stirring at room temperature. A drop of the suspension is placed on the surface of glass coverslips and allowed to dry for 15–20 min at room temperature to remove the excess of water. Therefore, the experiments are performed under a thin layer of water.

For AFM, a BioScope (Thermomicroscopes) or a BioProbe (NanoScope IIIa AFM, Digital Instruments) operating in contact mode is used in the experiments along with Si_3N_4 Nanotips (Digital Instruments) with 0.06-N/m spring constants. In some cases, the images are low-pass filtered to remove stray scan lines. All images are collected on the AFM using a scan speed of 2.5 Hz, and all imaging is performed in air at room temperature as described before.

Method

1. Stock solution of enzyme: 100 mg of enzyme is mixed in 1 mL doubly distilled and deionized water. Five microliters of this enzyme solution is spread on a 2-cm^2 area on a cover slips.
2. Building the nanobiosensor: a brand new cantilever is put in the cantilever holder of the AFM. An engage is given on the cover slips that contains the spread solution of the enzyme. After 30 s, the cantilever is taken off the cover slip surface and dried for 3 h at room temperature. The cantilever (with enzyme) is kept in the AFM during drying period, after which, this nanobiosensor is put in a vacuum desiccator for conservation under 5°C. To test the effectiveness of this methodology, an enzyme substrate solution is directly dropped on the nanobiosensor while it is scanning a glass surface; the reaction catalyzed by the immobilized enzyme is monitored by the deflections of the cantilever on the AFM monitor.
3. The nanobiosensors should be kept in a vacuum and under refrigeration. Under these conditions they can be used up to 2 or 3 mo after their preparation, and experiments with them are very reliable and reproducible.

22

Measurement of Single Molecular Interactions by Dynamic Force Microscopy

INTRODUCTION

Unbinding forces of weak, noncovalent bonds have been measured by scanning force microscopy or biomembrane force probes. Initially, these scanning force microscopy measurements focused on feasibility studies to measure single biomolecular interactions. Recently, however, a few groups showed that these single molecule experiments give a direct link to bulk experiments where thermodynamic data are experimentally acquired. In contrast with bulk experiments where averaged properties are measured, a single molecular approach gives access to properties that are hidden in the ensemble. These experiments can give insight into the geometry of the energy landscape of a biomolecular bond. Some experiments even showed that intermediate states during unbinding (unfolding) exist which only can be detected by single molecule experiments.

To understand the relation between force and energy landscape, one can consider an atomic force microscope (AFM) experiment in which the spring used to measure the forces acting on the molecular complex is weak compared with the molecular bond stiffness. The ligand is immobilized on a sharp tip attached to a microfabricated cantilever and the receptor is immobilized on a surface. When approaching the surface to the tip, a specific bond may form between ligand and receptor, e.g., complementary DNA strands or antibody-antigen. The bond is then loaded with an increasing force when retracting the surface from the tip (dynamic force spectroscopy; Fig. 1A).

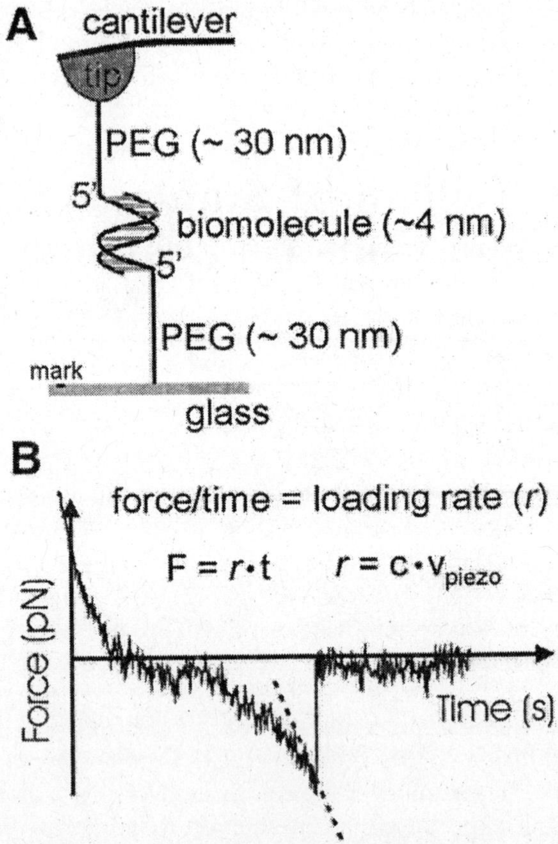

Fig. 1. (A) Single molecular interactions scheme measured by dynamic force microscopy. (B) Single biomolecule pulling experiment. The signal recorded is the force [pN] vs the time [s] (c = cantilever spring constant, v_{piezo} = pulling speed of the z piezo element). Keep in mind that the pulling of a biomolecule complex has to be performed at various pulling speeds in order to gather thermodynamic data. In the left part of the curve the cantilever is in its repulsive regime (above the time axis). As soon as the cantilever passes the equilibrium position and the biomolecular interaction has occurred, the cantilever is being deflected downwards. If a flexible cross-linker is fixing the biomolecule to the surface the linker is stretched as visible in the figure. At the point of rupture of the complex there is a sudden drop in force and the cantilever is ready for an additional pulling cycle. To extract the loading rate on the complex the derivative of the very last part of the force curve is fitted (dotted line).

At a certain force, ligand and receptor unbind, giving rise to an abrupt jump of the tip away from the surface (Fig. IB). It has been demonstrated that the unbinding is caused by thermal fluctuations rather than by a mechanical instability. If the thermal lifetime of the bond is short compared with the time it takes to build up an observable force during a slow loading process, no unbinding event is observed. With faster loading, finite unbinding forces are observed.

Therefore, unbinding forces depend on the rate of loading and on the details of the functional relationship between bond lifetime and an applied force. The theory for these experiments has been described in great detail in the articles from Strunz et al., Evans, and Merkel.

The technique has been applied to model systems such as biotin/(strept-)avi-din, antibody-antigen, or ssDNA-ssDNA where thermodynamic data existed.

But now because the link from single molecule experiments to thermodynamic ensemble experiments is clearly made the technique is applicable to other systems. The applicability of single molecule experiments for gathering thermodynamic properties now permits measurement in biomolecules in which the quantity of expressed molecules is barely suitable to allow a thermodynamic approach. In these systems (for example, the binding of a drug to specific receptors) the off rate (measurable by the single molecule techniques) of the ligand-receptor plays a key role for further studies or development.

In these days when more and more precious genetically engineered drugs are being developed and used, the storage of theses compounds is crucial.

If the compound is interacting with the surface of the storage container and no carrier substances like human serum albumin are allowed to be added, then the concentration of the substance in the containment is difficult to maintain during the shelf life of a drug. In a first study it was shown that force microscopy gives a suitable way to gather data from drug/storage container interactions and therefore to allow optimization of the storage container surface and buffer conditions in cases where storage problems occur.

MATERIALS

Conversion of the Reactive Groups on the Surfaces

1. Ultrasonic bath.
2. Strong ultraviolet (UV) light source (UV-Clean, Boekel Scientific, Feasterville, PA).
3. Argon.
4. Glass slides, thickness approx 0.4 mm, cut into 0.5 x 0.5 mm pieces.
5. Microcantilever, spring constant < 0.03 N/m.
6. Dry toluene, crown cap, molecular sieve.
7: Amino-propyl-triethoxy-silane (APTES).
8. Mercapto-propyl-triethoxy-silane (MPTES).

ACTIVATION OF THE REACTIVE GROUPS BY HETEROBIFUNCTIONAL

Cross-Linkers

1. Sulfosuccinimidyl 4-[A^-maleimidomethyl]-cyclohexane-1-carboxylate (SMCC) (Pierce, Rockford, IL).
2. Poly (ethylene glycol)-a-maleimide-<»-A-hydroxy-succinimide-ester (MAL-PEG-NHS), molecular weight 3400, length approx 30 nm (Shearwater Co., Huntsville; AL).
3. Dimethyl sulfoxide (DMSO).
4. Buffer, e.g., *N*-hydroxyethylpiperazine-*N*'-2-ethanesulfonate (HEPES), 2-(*N*-morpholino)ethanesulfonic acid (MES; free of primary amines) adjusted to pH 7.0–7.5, phosphate-buffered saline (PBS; pH 7.3; Life Technologies, Rockville, MD).

Coupling of the Biomolecules to the Activated Surface

1. Biomolecules (i.e., thiol-modified ssDNA, water-soluble proteins exposing free cysteins).
2. Ethyl-acetate.
3. Peltier element (Melcor, Trenton, NJ).
4. Thermocouple (Thermocoax, Suresnes, France).
5. Glove box (can be used but not mandatory).

METHODS

The methods described below outline (1) the conversion of the surface active groups to allow activation by heterobifunctional cross-linkers, (2) the activation of the newly generated groups by heterobifunctional cross-linkers, (3) the coupling of the selected biomolecules to the surface, (4) the set-up of the instrument to allow dynamic force spectroscopy, and (5) the extraction of the specific interaction parameters from the acquired data.

Conversion of the Reactive Groups on the Surfaces

The first steps of this procedure involve the cleaning of the individual surfaces (i.e., the flat glass surface and the microcantilever).

1. Mark two glass slides and clean the slides in ethanol for 10–20 min in an ultrasonic bath.
2. Dry under a stream of argon.
3. Replace ethanol in the beaker in the ultrasonic bath
4. Repeat this treatment twice.

   From now on, the glass slide surfaces, and the AFM-tips (e.g., Si_3Ni_4-Micro-cantilever, Park Scientific, Sunnyvale, CA), are treated in parallel.
5. Treat the surfaces with a strong UV light source for 60 min. Place the mark on the flat surfaces towards the UV light source.
6. Mixing of silanization solution: Insert one short syringe needle with an attached balloon containing argon through the crown cap (never remove the crown cap!). Take a syringe with a syringe needle long enough to reach the silanization solution (APTES). Withdraw enough solution from stock to enable a 2% dilution in *dry* toluene. Remove dry toluene comparable with two syringes through the crown cap.
7. Put the glass slides with the mark on the top and the AFM tips in a glass container that can be sealed. Immediately silanize the surfaces in a 2% solution of APTES in dry toluene for 2 h or overnight.
8. Rinse extensively with toluene and dry under a stream of argon.

5. *Activation of the Newly Generated Groups by Heterobifunctional*

Cross-Linkers

Immerse the surface in a 1-m*M* solution of MAL-PEG-NHS in DMSO for 2-3 h. The surfaces are again rinsed with DMSO and then with PBS buffer, pH 7.3.

Coupling of the Selected Biomolecules to the Surface

1. The oligonucleotides with a 57T-SH modification were synthesized by Microsynth (Balgach, Switzerland) and were stored in a PBS buffer at pH 7.3 containing 10 m*M* dithiothreitol (DTT) at 4°C until use. Immediately before use, the oligonucleotides are diluted to a final concentration of 25 m*M* with PBS buffer.
2. Extract DTT from an aliquot of typically 200 uL by washing three times with 1 mL of ethyl-acetate!
3. A 50-uL drop of the oligonucleotide solution is then incubated on the poly (ethyl-ene glycol)-co-maleimide-modified surfaces. Put the solution of one ssDNA oligonucleotide on top of flat glass slide (mark visible on the topside) and the complementary ssDNA oligonucleotide on the AFM-tips, which are on top of a piece of Teflon.
4. To avoid drying create a humidity box. Put some water bubbles on the wall of the box and close the box with a parafilm. Incubate overnight at room temperature in a humid chamber.
5. Rinse with PBS buffer, and then the tips and surfaces are ready for use in the force experiments.

Set-Up of the Instrument to Allow Dynamic Force Spectroscopy

Dynamic force spectroscopy measurements were performed using a commercial AFM instrument (Nanoscope IIIa, Digital Instruments, Santa Barbara, CA). The instrument has been expanded using the breakout box available from Digital Instruments. This allows an alternative control of the AFM by an external digital input-output board. We use the multifunctional DAQ board (PCI-MIO-16XE-10) from National Instruments

(Austin, TX) to have additional functionality to control the approach-retract function of the force microscope. The only functions of the Nanoscope used are the initial approach using the integrated stepper motor and the feedback for the first approach towards the sample. Additional features, which are controlled through the LabVIEW software package (National Instruments, Austin, TX), are:

Changing of the retract speed after each individual interaction. The speed is increased or decreased in exponential steps starting with speeds <10 pN/s and then increased up to approx 10000 pN/s. Normally we choose six different speeds within the speed-range of four orders of magnitude. The number of individual pulling cycles at a certain site can be defined before the location to another site on the sample is changed with nm precision.

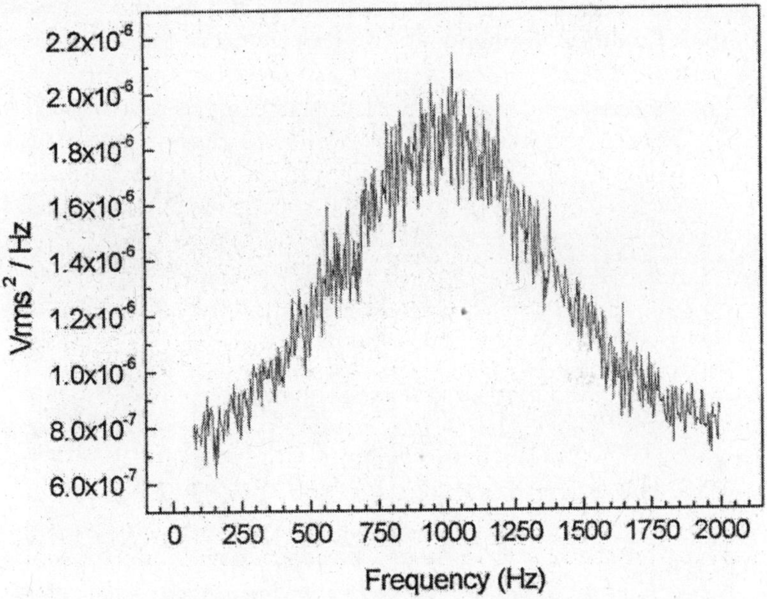

Fig. 2. The sampling rate is 4000 Hz, the frame size 1024 (step-size 3.90625 Hz), the number of points 512, spectrum shown is an average of 200 spectrums. The area under the curve is 0.00244 V_{rms}^2. With a sensitivity value of 0.0563V/nm, the spring constant is 5 pN/nm ± 10 % (compared with 10 pN/nm nominal value given by the company). The fitting and the integration of the volume can be done with commercial software (e.g., LABVIEW or Origin, OriginLab Corporation, Northampton, MA).

AFM cantilevers used for this experiment had spring constants <30 pN/nm. Each cantilever was *in situ* calibrated according to the method of Hutter. In short a power spectrum of the thermal vibrations of the cantilever is recorded using the LabVIEW software package. The spectrum shows a resonance peak at the natural resonance frequency. The integrated volume under the peak is directly correlated to the spring constant of the cantilever. It is important to retract the cantilever far enough from the surface (>1 μm) to avoid hydrodynamic damping. The temperature was controlled using a home-built fluid cell in which the buffer solution that immersed both the probe surface and the AFM cantilever was in contact with a Peltier element (Melcor, Trenton, NJ), driven with a constant current source. The temperature of the buffer was monitored with a thermocouple (Thermocoax, Suresnes, France). The thermocouple is calibrated with a digital thermometer and temperature measurements at different points of the cell showed deviations <2°C.

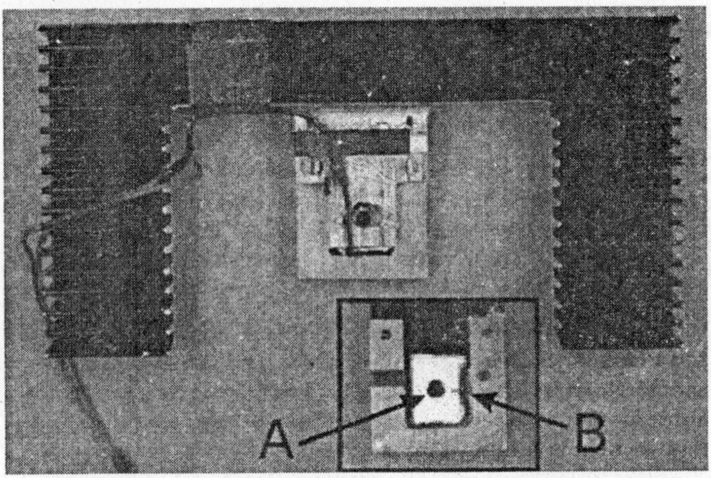

Fig. 3. Home-built fluid-cell platform allowing a precise temperature control. Big-picture complete device showing the top view of the cell including the big black heat sink block with lamellas for temperature stabilization. (Inset) Bottom view of the center of the fluid cell, which fits into the multimode head of the Nanoscope IIIa head. Arrow A, hole in the Peltier element for laser transmission. The hole is covered by a cover glass no. 1 to avoid fluid leaking and maintain angles of incidence of the laser. Arrow B, spring-to-clamp cantilever to the bottom of the Peltier element.

Extraction of the Specific Interaction Parameters from the Acquired Data

Write a small program to toggle through all the automatically collected force curves. The LABVIEW platform provides easy solutions, which can be expanded according to the operator's needs.

1. Select the force curves that show clear rupture force. Discard force curves that show no interaction or show unexpectedly long rupture distances.
2. Gather enough sample curves to obtain a reasonable number of data points to analyze your experiment. A way to analyze the data statistically is described by Izenman.

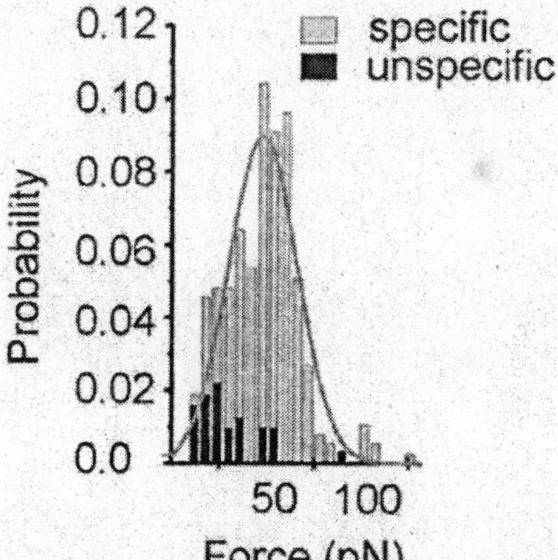

Fig. 4. ***Specificity check of the unbinding measurements. The force histogram can be fit with a gaussian distribution function to reveal the most-probable unbinding force (F*). Adjust the bin width according to Izenman. Gray bars, specific interaction; black bars, nonspecific interaction.***

3. Expand your force-curve analysis software to allow automatic recognition of the last jump back of the cantilever to the equilibrium position, determination of the slope at this point, the force of unbinding, and the speed of the pulling cycle (*see* Fig. 1B).
4. It is important to determine the specificity of the interaction measured. Use the second sample coated with the

unspecific biomolecules. Measure the interaction forces in the same way as with the 'specific' sample. You end up with a histogram comparable to Fig. 4. 5. Transfer your data into a diagram having the axes of F* (most probable unbinding force) and the rate of pulling. You should end up with a figure comparable to Fig. 5. Once you have collected and analyzed your data following our procedure you should be able to extrapolate the off rate of your system and all the other relevant parameters.

Notes

1. Plan to prepare the specific sample and the background sample in the same experiment. For instance we use two glass slides as substrates, one coated with the specifically interacting biomolecule (for example, ssDNA oligonucleotide) and the other with the unspecific molecule (for example, the nonhybridizing ssDNA oligonucleotide).

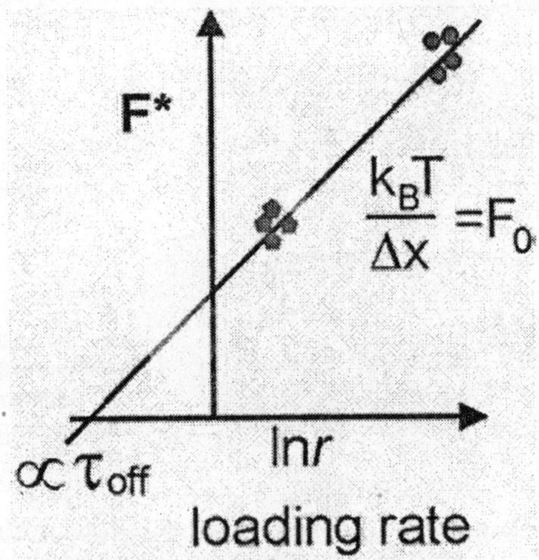

***Fig. 5.* Loading rate dependence of the unbinding measurements. The slope fit through the data sets provides insight into the energy landscape of the unbinding of the biomolecular interaction. If the precision of the data and the number of various pulling speeds is high enough to fit two slopes within this diagram then a probable intermediate state during unbinding can be detected.**

2. Silanization procedures provide an easy possibility to convert the surface reactive groups of silicon or glass or

even silicon nitride tips of the cantilever. Remember that most groups have their own ideal preparation procedure. The activation of surfaces with trifunctional (for example, tri-ethoxy) silans easily can lead to monolayers on the surface of the sample (*see* Fig. 6), but there is a great chance that multilayer and aggregates of polymerized silans are formed (*see* Fig. 7). To reduce this possibility monofunctional silans can be used. The reactivity of these substances is comparable but one should work in an argon environment during the whole silanization procedure (glove box). Curing the surface by heating the sample (150°C) after silanization is possible but not necessary in the case of trifunctional silans.

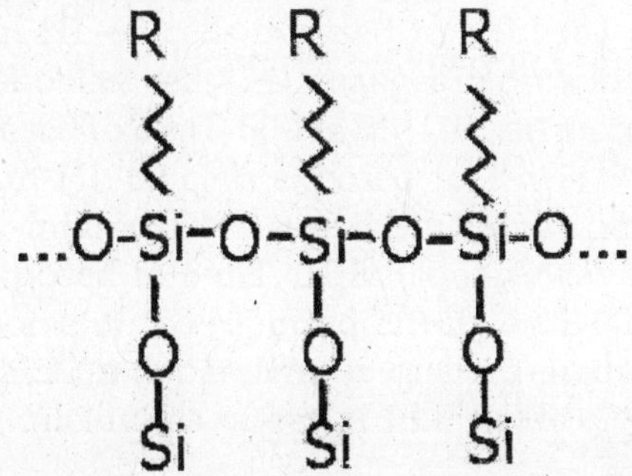

Fig. 6. Ideal surface activation (silanization).

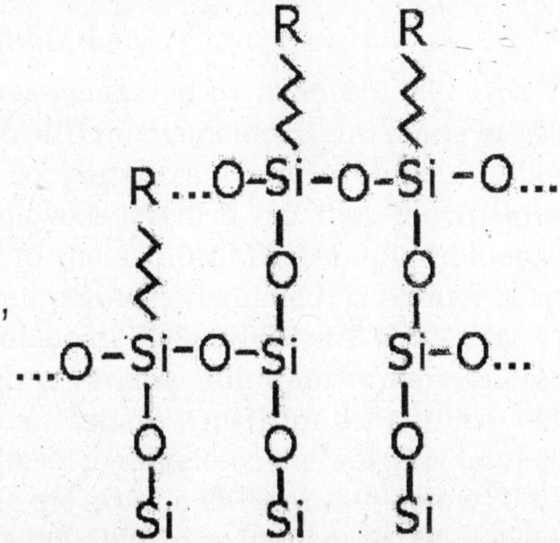

Fig. 7. Reality: partial hydrolysis during deposition, multiplayer formation, and formation of aggregates.

3. Long flexible linkers are needed when single biomolecule interactions are being probed. When manipulating single globular proteins, tertiary folded RNAs or ssDNA–ssDNA pairs, additional problems must be overcome *(19,20)*. As a result of their small dimensions (typically 1.5–6 nm) direct tethering to the surfaces is undes-ired. Furthermore, the nonspecific interactions of the surrounding surfaces would mask the signal of the small individual molecule. These interactions, and possible solvent exclusion within the gap, could lead to denaturation of the molecule or perturbation in the force profiles. A covalent attachment of these biomolecules to surfaces via long linkers minimizes the nonspecific interaction of the surfaces with the biomolecule, and allows probing the single protein molecules. Long linkers of some tens of nanometers from the surface allow space for free exchange of additional ligands or solution in the environment of the molecular interaction being probed. Additionally the use of long linkers allows easy discrimination of single events from multi-events. In cases were two molecular interactions happen a saw-tooth-like pattern in the retract curve is visible. The probability that two bonds break simultaneously is minimized because the length of the flexible cross-linker is comparable to the curving of the cantilever tip. Because the anchoring points are located on a highly curved surface, simultaneous ruptures are very unlikely.
4. In experiments were short linkers are applied the surfaces are immersed in a 1 m*M* solution of sulfo-SMCC in 50 m*M* HEPES buffer (pH 7.0) for 1 h at room temperature. After incubation rinse the surfaces with HEPES buffer only to remove excess of cross-linkers. Short linkage of the ligand/ receptor system on the surface is resulting in considerably more unspecific adhesion events. It is still possible to extract the "real" value of the unbinding force if the very last jump back to equilibrium is being considered. Important is the change in pulling speed during the experiment to obtain off rates and not just one interaction force. Owing to the fact that multiple binding events can rupture at the

point of rupture a bimo-dal or higher modal distribution in the force histogram is to be expected

5. It is important to note that the activated cantilever surface has a limited lifetime and therefore the speed of the following steps is crucial. The cantilever is thoroughly rinsed with the buffer used to store the biomolecule. Because the specificity of the cross-linker is changing with the pH and is sensitive towards free amino groups. Buffers like Tris-HCl should be avoided. Buffers of choice are PBS, HEPES, and MES, adjusted to the right pH. The thiol groups of the biomolecules are coupled at a pH ranging from 7.0–7.5 at room temperature.
6. Biomolecules have to be cleaned if stored under protective conditions like those mentioned for thiol-modified oligonucleotide (DTT has been added to prevent oxidation of the thiol to the disulfide). Extraction with organic solvent is an easy way to remove these compounds. By repeating the extraction the amount of DTT is reduced to levels that do not interfere with the covalent coupling to the maleimide group. Oligonucleotides remain in the water phase and DTT is transferred in the organic phase (top phase of the two).
7. In addition, be careful with the buffers in which the biomolecules are being delivered. Unsuitable buffers or protective agents quench the coupling reaction completely. These contaminants are removed using centrifugation membranes available from various companies (e.g., Amicon, Eppendorf, Sartorius). Calculate the dilution factor to be applied during this cleaning to bring the 'contaminant' to a level of approx 1/100 relative to the molecule to be coupled.
8. If instead of thiol groups of the biomolecule (thiol-modification on ssDNA end or free cysteine of protein) amino groups (amino-modification on ssDNA end or free lysine or terminal amino group on protein) have to be coupled to the surface, then instead of APTES, the silanization reagent MPTES can be applied and the protocol follows exactly the same description.

9. The cross-linking group maleimide has its optimal pH range from 6.5–7.5 and succinimide from 7.0–8.5. Some adjustments of pH during the individual coupling steps of the individual groups can improve the overall performance of the coupling reaction.
10. To calculate the spring constant of the cantilever and to express the cantilever deflection as a force, the cantilever deflection signal vs the voltage applied to the piezo tube (sensitivity) has to be measured. It is equal to the slope of the force curve while the cantilever is in contact with the sample surface (a cantilever deflection signal of 1–2 V on the photodiode is sufficient for obtaining a good value). Typically, with a spring constant of approximately 10 pN/nm, the sensitivity values are around 0.07–0.09 V/nm. So 1 V deflection signal on the photodiode corresponds to 110–150 pN.

Bibliography

Arend, A.: Introduction to Atomic Force Microscope, London, Routledge, 1993.

Bellamy, C.: Knights in White Armour: The New Art of War and Peace, London, Random House, 1997.

Damrosch, L.F.: Human Fibroblasts in the New International Order, Boulder, Westview Press, 1991.

Friedman, G.: The Future of War: Power, Technology and American World Dominance in the 21st Century, N.Y., Random House, 1996.

Greer P.: Transforming Central Government: the Next Steps Initiative, Buckingham, Open University Press, 1994.

Huntington, S.P.: The Soldier and the State, N.Y., Vintage Books, 1964.

Janowitz, M.: The Professional Soldier: Cardiac Physiology, N.Y., The Free Press, 1960.

Mary, W.: Military Space Force: the Cultural Dimensions of Authority, Cambridge, Belknap Press, 1985.

Posen, B.R.: The Sources of Military Doctrine, Ithaca, 1984, Cornell Univ. Press.

Peterson, B.R.: The Sources of Military Space Force, Cornell Univ. Press.

Rosen, P.: Societies and Military Power: India and its Armies, Ithaca, Cornell University Press, 1996.

Index

□□□